Introduction to Structural Aluminium Design

Ulrich Müller

Whittles Publishing

Published by
Whittles Publishing,
Dunbeath,
Caithness KW6 6EG,
Scotland, UK

www.whittlespublishing.com

Distributed in North America by
CRC Press LLC,
Taylor and Francis Group,
6000 Broken Sound Parkway NW, Suite 300,
Boca Raton, FL 33487, USA

ISBN 978–184995–007–7
USA ISBN 978–1–4398–5468–6

The publisher and author have used their best efforts in preparing
this book, but assume no responsibility for any injury and/or damage to persons or
property from the use or implementation of any
methods, instructions, ideas or materials contained within this book.
All operations should be undertaken in accordance with existing
legislation and recognized trade practice. Whilst the information
and advice in this book is believed to be true and accurate at the time
of going to press, the author and publisher accept no legal responsibility
or liability for errors or omissions that may have been made.

Cover image of the 'Glasgow Armadillo' used with the kind permission of
Alan Moir (the bloofish.com)

Permission to reproduce extracts from Eurocode Basis of Structural Design
and Eurocode 9 is granted by BSI. British Standards can be obtained in PDF or
hard copy formats from the BSI online shop: www.bsigroup.com/Shop or by
contacting BSI Customer Services for hardcopies only: Tel: +44 (0)20 8996 9001,
Email: cservices@bsigroup.com

Typeset by
iPLUS Knowledge Solutions Private Limited, Chennai-32, India.

Printed by
Bell & Bain Ltd., Glasgow

Contents

Preface

The aim of this book is to highlight the properties and applications of aluminium alloy members. But this is not limited to their strengths. It aims to identify not only their strengths but also their weaknesses and possible work-arounds. It is the author's intention to give basic, but sufficient, advice and information about the material itself, fabrication, structural design and corrosion concerns.

Working with aluminium elements in the construction industry for many years has shown that both engineers and architects have only ever manifested a slight interest in using aluminium as a structural material. Whilst the benefits of aluminium alloys have been utilised in the automotive and transport industry, in construction aluminium elements have been used mainly for curtain walling and in association with patent glazing. With structural strength properties close to steel, except for the lesser modulus of elasticity, and taking the special properties of aluminium alloys into account, it should be possible to make better use of this excellent material.

When designing structures made from aluminium alloy sections, there was a noticeable tendency for there to frequently be questions and queries about the lesser modulus of elasticity, indicating that aluminium alloy sections were considered to be weak or soft.

During innumerable site visits and inspections a lack of understanding of the peculiarities that aluminium alloys possess was noticeable. Often laborious remedial work was required to obliterate errors made in the design, fabrication and protection of the structures that were being inspected.

With the introduction of the new Eurocodes in 2010, the design methods of EN 1999 (also called Eurocode 9), will be discussed and illustrated to describe these procedures when designing structural elements using aluminium alloy members. However, the main objective is not simply to produce a design guide for structural members made from aluminium alloys, it is rather to examine and discuss the use of the material for structural design as such, from production to fabrication.

This book aims to present a range of subjects associated with the structural design of aluminium elements and to consider the main aspects such as the material itself, structural analysis, serviceability, element design and fatigue, all of which relate to the structural design of aluminium members.

The author wishes to offer concise information for students and engineers, but also anyone else who has an interest in the material aluminium, the use of aluminium for structural and non-structural applications and an introduction to designing structures made from aluminium or aluminium alloy elements.

Hopefully readers will find this book of use, interesting and inspiring.

Ulrich Müller

Acknowledgements

Grateful acknowledgement by the author is made to all supporters, who contributed to the development of this publication.

The author wishes to thank:
   Aldel Delfzijl BV, Delfzijl, the Netherlands;
   Brital Limited (Aluminium Façade Designs), Walsall, UK;
   the Aluminium Federation Ltd (ALFED), West Bromwich, UK;
   the Aluminium Centrum Amsterdam, the Netherlands;
   and the International Aluminium Institute (IAI), London, UK for
      permission to include their pictures and graphs in this book.

The author would also like to thank the IAI, London for providing
   statistical data and graphical resources.

1

INTRODUCTION

Whilst structural steel, timber, masonry and concrete are widely known and used in the engineering community, and sufficient design guidance and structural design software is widely available, structural aluminium elements are used more sparingly. That they are overshadowed by these other materials is unjustified and might be based on a lack of experience in the use and design of aluminium structural elements. For engineers, the amended British Standards, new European Design Codes and missing guidance for structural aluminium, generates difficulties in knowing and applying the necessary procedures when designing aluminium elements.

Aluminium, as aluminium alloys, has been extensively used in the aviation and transportation industries where the lightweight material is appreciated for its ability to reduce the self-weight of planes and vehicles, resulting in increased loading capacity. Aluminium alloys have been utilised for curtain walling, patent glazing and manufacture of windows and doors, where the ease of fabrication but also an infinite range of possible section shapes due to extrusion techniques are among its main advantages.

The pure and soft aluminium must be strengthened by alloying to achieve good mechanical properties, an enhancement process which results in tensile strengths for up to 500 N/mm^2, comparable to the properties of quality steel.

However, the good tensile strengths which can be achieved cannot fully compensate for the lesser modulus of elasticity, the fact of material weakening adjacent to welds (heat-affected zones (HAZs) buckling and the susceptibility to fatigue. Yet, these limitations of aluminium alloys can be designed out. Suitable workarounds can be used to greatly reduce the weak spots and provide the required structural strength and cross-sectional properties.

Smart design, exploiting the strengths of the material in tension and compression, limiting the exposure to bending actions, will result in structures with no less structural integrity than the same structure

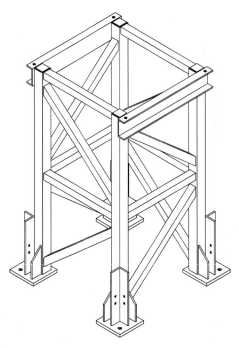

Figure 1.1 | Typical lattice structure

made of mild steel. A good example for this practice is the use of lattice-type structures (see Figure 1.1).

Fabricated from aluminium alloy sections, this structure demonstrates structural capacities equal to a mild steel structure, however, due to the lightweight material it is possible for one person to carry this prefabricated structure. Utilising hollow sections with short unrestrained lengths reduces exposure to bending and lateral-torsional buckling, resulting mostly in compression and tension members. Nevertheless, allowance has to be made for HAZs. Post-fabrication treatment has to be applied to reduce susceptibility to fatigue adjacent to welds. Whilst the structure itself will not require protective treatment against corrosion, the bases must be treated to prevent corrosion as they are in contact with other materials, in this particular case concrete and contaminated rainwater which sprays over the concrete base.

Aluminium alloys are often praised for their excellent resistance to corrosion in most environments. Nevertheless, various types of corrosion are possible, where damage to the surface protective layer can occur due to weathering, imperfections and bi-metallic reactions. Knowledge of these corrosion concerns will allow suitable measures to be used to protect the members from exposure to such attacks.

With sustainability and recyclability requirements introduced to building and construction, the use of recycled aluminium can certainly help to achieve the targets that have been set and limit the impact on the environment. Whilst the primary production of aluminium consumes large quantities of energy, re-smelting aluminium requires only approximately 5% of the energy used for primary production. When combining the energy saving aspect of recycling with the unique property of aluminium to be infinitely recyclable without loss of quality recycling, makes aluminium a good contestant for true sustainability in construction. The ease with which aluminium can be recycled makes it a simple and cost-effective process. Once any contaminants have been removed the material is re-smelted and cast into new ingots. Admittedly, the environmental effects of collection, transportation and re-smelting still cast a cloud over the otherwise very positive development of aluminium recycling. The use of renewable and low-carbon energies for the recycling process could further enhance its claim to be a sustainable material.

Aluminium products are available in various shapes and section types, but are often delivered semi-finished and require further fabrication to produce the required end shape. Fabrication methods are also necessary to connect the members forming the structure or frame. These fabrication techniques such as: sawing, cutting, drilling, punching, machining, forming and jointing, also have their advantages and disadvantages. The softness of aluminium makes it easy to machine and form. It is easier to cut, saw, drill and punch than steel products, thus making it a preferred material for the fabricator. Nevertheless, the analysis of the fabrication methods yielded weaknesses and imperfections of the material such as: weakening adjacent to welding, vulnerability to heat and spring-back when bent, to name a few only. Not all aluminium alloys are weldable or react positively to heat treatment. Knowledge about the available alloys and tempers and their specific properties is critical for material selection, a major part of the overall structural design.

Despite those material properties that allow an ease of fabrication, similar to but easier than that of steel fabrication, many aluminium alloys develop weakened zones when welded. The heat generated by the welding process reduces the strength of the material in the vicinity of the welds. The zones affected by weakening due to welding are described as HAZs. HAZ softening can only be ignored for material supplied in the annealed or T4 (see Section 2.7) condition. For all other alloys and associated tempers allowance for this material weakness must be made in structural design. At first glance HAZ softening appears to be a major drawback for the designer of aluminium

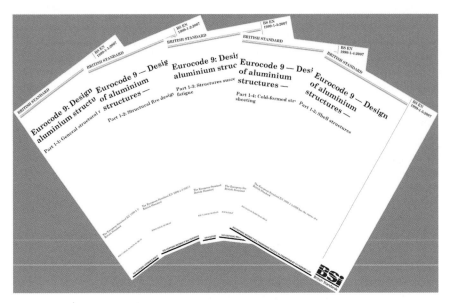

Figure 1.2 | Eurocode 9 family

structures. However, measures to limit welding to the absolute minimum, together with moving welds to less stressed locations can often achieve the desired result.

Eurocode 9 for the design of structural aluminium harmonises the design procedures used for structural design of aluminium alloys. With a familiar section classification and design checks throughout the entire Eurocode family (see Figure 1.2) structural aluminium design procedures should no longer cause engineers and architects to have reservations.

At the present, the trading prices for aluminium and aluminium alloys are still high and approximately three times the trading price for steel. This is only offset by the weight of aluminium and aluminium alloys at more or less one-third that of mild steel. With raising trading prices for steel and increasing rates for recycling aluminium, this development could benefit the use of aluminium alloy structures in the future.

2

ALUMINIUM: THE MATERIAL

Introduction

The rising price for structural steel in the UK may vitalise the use of structural aluminium as the economic advantage of structural steel over aluminium diminishes. In theory weight savings up to 70% compared to steel and up to 90% compared with reinforced concrete construction might be achievable. As a result, aluminium has the prospect of becoming the cheaper option in comparison to steel and concrete structures. Whether or not such weight savings are actually possible depends on the load ratio of dead load to imposed load. The higher the dead load/imposed load ratio, the more weight saving can be expected. Thus, long-span constructions with high dead load/imposed load ratio are apparent candidates for the cost-efficient structural use of aluminium materials.

Aluminium has a bright future as a structural material: its respectable structural properties, excellent resistance to corrosion, being lightweight and fully recyclable, together with a good understanding of the properties of aluminium alloys make it a good structural material.

2.1 History

In 1807 the British scientist Sir Humphrey Davy suggested the existence of the element 'alumium'. Sir Humphrey Davy was a Cornish chemist and physicist and at the time he was working on the separation of salts by electrolysis. In 1821 the French geologist Pierre Berthier discovered a hard, reddish, clay-like material containing 52% aluminium which he named bauxite. It is the most common ore of aluminium. Later in 1825, the Danish chemist and physicist Hans Christian Ørsted produced metallic aluminium for the first time, by heating potassium amalgam with aluminium chloride. The German chemist Friedrich Wöhler also attempted to extract the metal aluminium and co-discovered aluminium in 1828. By 1845 he had established many of the new metal's properties. In 1855, the French chemist, Henri Etienne Sainte-Claire Deville succeeded in obtaining metallic aluminium by a reduction process using sodium. This was the first attempt to produce metallic aluminium in commercial quantities. In 1886 Charles Martin Hall, an American inventor and engineer, developed an inexpensive method to produce aluminium. At the same time, Paul (Louis-Toussaint) Héroult discovered the electrolytic aluminium process (see Figure 2.1). The process was later named the Hall–Héroult process, and 1888 Hall opened the first large-scale aluminium production plant in Pittsburgh, Pennsylvania. In 1887–1889 Karl Josef Bayer, the son of Friedrich Bayer (founder of Bayer Chemical and Pharmaceutical Company), invented the Bayer process for extracting alumina from bauxite, a process that is still fundamental to today's production of aluminium. Industrial production of aluminium started in Switzerland and France in 1888.

The first aluminium alloy designation dates back to 1888 and since then a vast number of aluminium alloys have been developed until the present day. In 1903, Alfred Wilm, a German metallurgist, discovered an aluminium alloy containing 4% copper that after quenching would harden slowly if left in room temperature and 'Duralumin' was introduced in 1909 by the Dürener Metallwerke Aktien Gesellschaft in Germany. And by 1912 the science of alloying has been widely established. The non-heat-treatable (NHT) alloys were the next to be developed. In the 1920s, Birmetals Co. produced and marketed 'Birmabright'

Figure 2.1 │ First Héroult Cells in 1889 (Reproduced with permission of Aluminium Federation, Ltd, West Bromwich, UK)

an alloy of aluminium and magnesium with extra good corrosion resistance. By the end of 1939 most of today's alloys had been developed. The weldable 7xxx type were introduced after the Second World War.

Key dates:	
1807	First mentioned by Sir Humphrey Davy
1821	Pierre Berthier discovered bauxite
1825	First production by Hans Christian Ørsted
1828	Extraction of metallic aluminium by Friedrich Wöhler
1845	Friedrich Wöhler established density of aluminium
1855	Henri Etienne Sainte-Claire Deville obtained aluminium
1886	New methods by Charles Martin Hall and Paul L T Héroult
1888	First aluminium companies in France, Switzerland and USA
1889	Karl Josef Bayer invented the Bayer process
1900	Annual aluminium production 8,000 tonnes
1909	Duralumin introduced by Alfred Wilm
1939	Availability of all alloys, except weldable 7xxx-series

Since the first commercial production of alumina the annual manufacture of alumina and aluminium has increased steadily, illustrating the growing popularity of aluminium products across all industries. The production of aluminium increased from less than 200 tonnes in 1885, to approximately 25 million tonnes in 2008. At the start of the 21st century, the main producers of aluminium are Europe and North America.

World production of alumina increased again comparing data from the year 2007 to the year 2008 (see Figure 2.2). Based on production

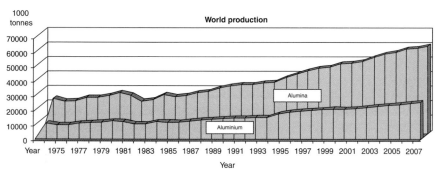

Figure 2.2 World production of alumina and aluminium (data supplied by IAI, London, UK)

data from the International Aluminium Institute, world alumina production during the first two quarters of 2006 increased 9% compared with that for the same period in 2005. However, in common with all natural assets, bauxite resources are limited. Seen on a global scale, bauxite resources are estimated to be 55–75 billion tonnes, located in South America with 33%, Africa with 27%, Asia with 17%, Oceania with 13%, and elsewhere with 10%.

2.2 Manufacture

2.2.1 Primary production of aluminium metal

The production of aluminium ingots involves three main processes: mining the bauxite ore, refining of bauxite to gain alumina, and smelting of alumina to extract aluminium (see Figure 2.3).

Bauxite ore consists of the minerals gibbsite [$Al(OH)_3$], böhmite/diaspore [$AlO(OH)$], the iron oxides goethite [$FeO(OH)$] and hermatite [Fe_2O_3], the clay mineral kaolinite [$Al_2Si_2O_5(OH)_4$] and anatase [TiO_2]. It is estimated that approximately 90–95% of the entire amount of bauxite that is mined is processed into aluminium. Today most bauxite is mined in Australia, West Africa, Brazil and Jamaica. It is crushed, dried and shipped to the alumina plants in Europe and America, but also Asia. Approximately 2 kg of bauxite are required to give 1 kg of alumina.

Bauxite is refined to obtain alumina by the Bayer process (see Figure 2.4), where bauxite is washed, milled and dissolved in sodium hydroxide [$NaOH$] at a high temperature. The accrued fluid contains a solution of sodium aluminate and undissolved bauxite

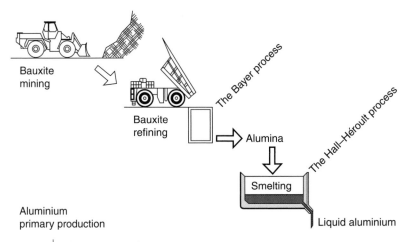

Figure 2.3 | Aluminium production

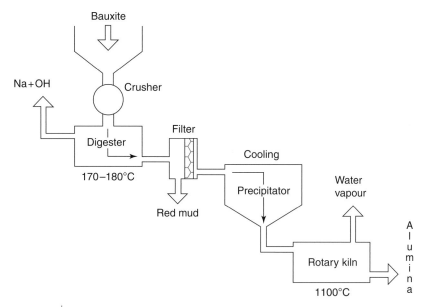

Figure 2.4 │ Principle of Bayer process

deposits. These undissolved residues, known as red mud, sink to the bottom of the digester tank and are filtered and removed. The left-over sodium aluminate [$Na_2Al_2O_4$] is pumped into the next tank, known as a precipitator. During the cooling process fine particles of pure alumina sink to the bottom of the precipitator. The pure alumina particles are removed and passed through a rotary kiln at very high temperature to free and remove chemically bound water. The end product is a white powder, pure alumina [Al_2O_3] also known as aluminium oxide.

Metallic aluminium is extracted from alumina powder by an electrolytic procedure, named after Charles Martin Hall and Paul L T Héroult, the Hall–Héroult process. It is still the only method of aluminium smelting used by aluminium producers around the world.

Alumina is dissolved in a carbon-lined steel container in molten cryolite [Na_3AlF_6] and aluminium fluoride [AlF_3]. Aluminium fluoride is used to lower the melting point of cryolite. The mixture is electrolysed using low voltage at approximate 3–5 V and high electrical current at around 150,000 A. The liquid aluminium is drawn to the cathode and deposited. The metal forms at a temperature of 900°C and has a very high purity of 99.5%. As the liquid aluminium has a higher density as molten cryolite it sinks to the bottom of the container and is removed at regular intervals. The carbon anodes require frequent replacement and to produce 1 kg of aluminium, 2 kg alumina, 0.5 kg carbon and 15 kWh of electrical energy are needed. The smelting process is continuous and cannot easily be stopped and restarted (see Figure 2.5).

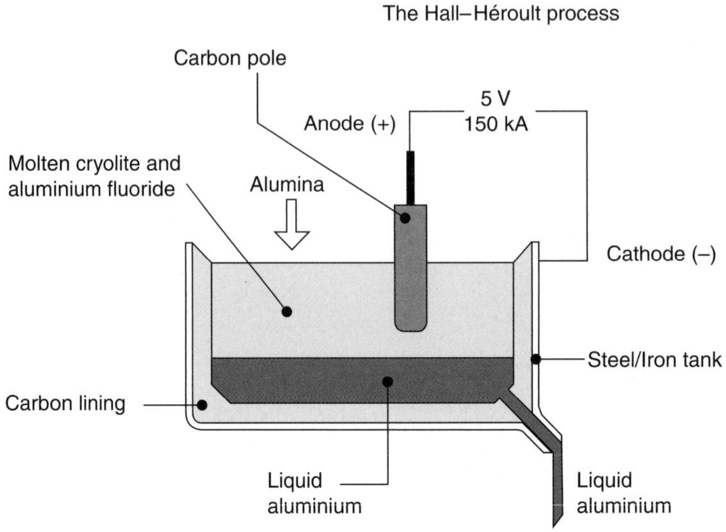

The Hall–Héroult process

Carbon pole

Anode (+)

5 V
150 kA

Molten cryolite and
aluminium fluoride

Alumina

Cathode (–)

Steel/Iron tank

Carbon lining

Liquid
aluminium

Liquid
aluminium

Figure 2.5 | Principle of Hall–Héroult process

If production is interrupted long enough, the metal in the pots will solidify, requiring an expensive rebuilding process.

A modern aluminium plant can produce annually up to 250,000 tonnes aluminium with one smelter. One smelter typically consists of 300–500 'pots'. At the primary smelter the metal can be cast into ingots or larger blocks for subsequent re-melting. The metal is usually alloyed at the smelter and then cast into cylindrical extrusion billets or rolling ingots in form of rectangular slabs. These primary aluminium products are then shipped to aluminium factories where they are processed further into semi-fabricated products. Semi-fabricated products are produced by the processes of rolling, forging, casting and extrusion. Figure 2.6 shows a typical ingot cast for shipping to processing factories.

2.2.2 Secondary production, recycling of aluminium metal

All aluminium is recyclable after use. As reprocessing does not damage the metal or its structure, aluminium can be repeatedly recycled. Once taken to the recycling plant, aluminium scrap will be checked and sorted to determine the composition and value of the scrap metal. Depending on the type of contamination some aluminium scrap must be processed further to remove coatings and other contaminations. The 'clean' aluminium scrap is then melted in furnaces. Subsequently the molten aluminium metal is cast or processed using the same methods as for primary processing.

Figure 2.6 │ Aluminium ingot as produced (Reproduced with permission from Aluminium Centrum Amsterdam, the Netherlands and Aldel Delfzijl BV, Delfzijl, the Netherlands)

The recycling of aluminium scrap to produce aluminium metal requires only approximately 5% of the energy used to produce the metal in primary production. And recycling of scrap aluminium saves approximately 6–8 kg of bauxite and 4 kg of other chemicals. Figure 2.7 illustrates the high recycling rate in aluminium production where the secondary aluminium production accounts for more than 1/3 of the total production of aluminium.

Due to the high demand for scrap aluminium and scrap metal prices continuously rising, the value of scrap aluminium (see Figure 2.8) has attracted criminal activity in recent years. Police and press recently

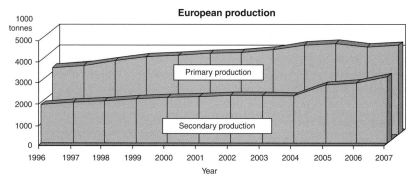

Figure 2.7 │ European primary and secondary production (Data supplied by IAI, London, UK)

Figure 2.8 | Aluminium scrap metal (Reproduced with permission of IAI, London, UK)

reported 'gang' activity and record highs in thefts of metal. The metal theft is not limited to scrap metal but includes the removal of manhole covers in streets, cutting down aluminium posts and signs along highways, demounting of rainwater down pipes and gutters and even break-ins stealing wiring materials.

2.3 Physical properties

2.3.1 Properties of pure aluminium

Aluminium is a chemical element and has the symbol Al (see Figure 2.9). In the periodic table it has the atomic number 13 and has an atomic weight of approximate 27 g/mol. The nucleus of the aluminium atom contains 14 neutrons and 13 protons.

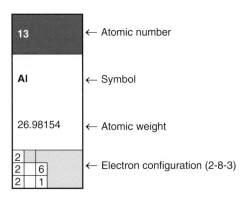

Figure 2.9 | Extract from periodic table

Pure aluminium is a soft, lightweight metal with a light-grey silvery appearance. This is due to the oxidation that forms a thin layer when aluminium is exposed to air. Aluminium, the metal is non-toxic, non-magnetic and non-sparking. Due to the process of passivation, aluminium has a very good resistance to corrosion.

Aluminium is found primarily as bauxite ore and is the third most prevalent element in the earth's shell. It is the most plenteous metal and count for roughly 8% of the total mass of the earth's crust.

Aluminium is non-combustible and its melting point is 933K (660°C). Aluminium loses strength and distorts when overheated. It will only burn when first finely divided into a powder, where it will burn with no colouration of the flame.

Pure aluminium is classified as a weak metal with a tensile strength in the range 90–140 N/mm². In its pure form, aluminium is not only used for domestic products such as cans and packaging but also for electrical conductors. Thus, aluminium for structural applications must be strengthened. This strengthening is achieved by the method of alloying. In this way, tensile strengths of in the region of 500 N/mm² have been reached. Table 2.1 summarises the main physical properties of aluminium.

Table 2.1 | Properties of pure aluminium

Property	Value	Notes
Atomic Number	13	
Atomic Volume	10 cm³/g-atom	
Atomic Weight	26.98	
Coeff of Thermal Expansion	$\alpha = 23.5 \times 10^{-6}/ °C$	$\alpha_{200} = 26 \times 10^{-6}/ °C$ at 200°
Density	$\rho \sim 2.7$ g/cm³	~ 2700 kg/m³ at 20 °C
Electrical Resistivity	$R = 2.69 - 2.824 \ \mu\Omega cm$	At 20°C ambient temperature
Elongation	~50%	Maximal before breaking
Hardness	BHN = 15 Brinell	
Modulus of Elasticity	E = 69 kN/mm²	$E_{100} = 67$ kN/mm² at 100°C $E_{200} = 59$ kN/mm² at 200°C
Modulus of Rigidity	G = 26 kN/mm²	$G = E / 2(1 + v)$
Point of Melting	~660° C	

(Continued)

Table 2.1 | *(Continued)*

Property	Value	Notes
Point of *Boiling*	~1800 – 2480° C	
Poisson's Ratio	$v = 0.33$	Baker and Roderick,1948
Proof / Yield Stress	$f_y = < 25 N/mm^2$	
Thermal conductivity	$K = 240$ W /m° C	At 20°C ambient temperature
Ultimate Tensile Strength	$f_{y\,ult} = < 58$ N/mm²	
Specific Heat	$c = 22$ cal/g° C	
Valency	3	

2.3.2 Uses or applications of aluminium

Lightweight: The weight of aluminium is only one-third that of steel. This property is important for the transportation industries building aircrafts, trains, ships but also cars. It is appreciated in high-rise building and bridges to limit the self-weight of large structures.

Corrosion resistance: Exposed to air, aluminium forms its own protective layer against corrosion. Anodisation further increases the corrosion resistance of aluminium. This is utilised in the transportation industry, construction industry and manufacture of household tools and appliances.

Ease of fabrication: Due to its softness, aluminium can be fabricated into various forms and shapes. It can be machined with ease and good plasticity allows bending, cutting and drawing. It is considered best material for complex-sectioned extrusions.

Non-toxic: Aluminium is non-toxic and odourless. Its smooth metallic surface can easily be cleaned and kept hygienic. It is often used to produce food cans and packaging, cooking utensils and in the food industry.

Non-magnetic: With its non-magnetic attribute, aluminium finds use in magnetic compasses, antennas and computer discs.

Electric conductivity: Aluminium is widely used as an electric conductor in power transmission cables (where the weight is also beneficial), substation conductors but also for the bases of electric bulbs.

Heat conductivity: Aluminium is approximately three times as thermally conductive as steel. This is employed for cooking equipment, air-conditioners, heat exchangers, engine parts and solar collectors.

Low temperature use: Unlike steel, aluminium increases in tensile strength at low temperatures, retains good quality and remains non-brittle. Therefore, aluminium structures can be used in extreme environmental conditions.

Reflectivity: Aluminium is highly reflective of light, heat and electric waves. This characteristic is applied for mirrors, heat reflectors, insulation, lighting equipment and waveguides.

Recyclability: Due to the low melting point, aluminium can be recycled easily and economically. Thus, it is a modern material, energy and resource saving.

2.4 Aluminium and its alloys

2.4.1 Designation and numbering of aluminium alloys

Pure aluminium is a weak material with little strength for structural engineering applications. However, pure aluminium can be strengthened by alloying and subsequent treatment. Aluminium alloys are grouped in wrought aluminium alloys and cast aluminium alloys and divided into eight alloy series 1xxxx–8xxxx. An additional differentiation is made by subdivision into heat-treatable (HT) and non-heat-treatable (NHT) alloys. Wrought aluminium alloys are more commonly used as cast alloys contain a higher percentage of alloying elements and due to manufacturing defects have reduced strength properties. The most commonly used alloying elements are copper, magnesium, manganese, lithium and zinc. Based on the main alloying elements used, aluminium alloys are numbered and incorporated within the designated alloy series.

Guidance on the aluminium alloy designation can be found in the British Standards BS EN 573 for wrought aluminium alloys and in BS EN 1780 for cast aluminium alloys. Tables 2.2–2.4 give a summary of the numerical designation system used.

The current aluminium alloy designation system for AWs is based on the well-established system for alloy designation administered by the USA-based Aluminum Association (AA) Inc. Most international aluminium associations have signed to use the numerical designation system for AWs known as Teal Sheets. These sheets give the relevant designation and also chemical composition of the alloy. The Teal Sheets can be obtained from the AA Inc., Arlington, Virginia, USA or downloaded via their website.

The first digit in the designation relates to the alloy group, as indicated in Table 2.2, and is directly associated to the major alloying

Table 2.2 | Numerical designation system for wrought aluminium alloys (AWs)

Series	Alloying elements	Type
1xxx	None (aluminium 99% and greater)	NHT
2xxx	Copper (Cu)	HT
3xxx	Manganese (Mn)	NHT
4xxx	Silicon (Si)	NHT
5xxx	Magnesium (Mg)	NHT
6xxx	Magnesium and silicon (MgSi)	HT
7xxx	Zinc (Zn)	HT
8xxx	Other elements	

HT = heat-treatable, NHT = non-heat-treatable

Table 2.3 | Numerical designation system for ACs

Series	Alloying elements
1xxxx	None (aluminium 99% and greater)
2xxxx	Copper (Cu)
3xxxx	n/a
4xxxx	Silicon (Si)
5xxxx	Magnesium (Mg)
6xxxx	n/a
7xxxx	Zinc (Zn)
8xxxx	Tin (Sn)
9xxxx	Master alloys

AC = Cast Aluminium Alloys

element used. The second digit designates a modification from the specific alloy unless this digit is 0. The last two digits identify the alloy in its series. For the pure aluminium alloys in the alloy group 1xxx, the last two digits give the percentage of purity above 99%, for example the alloy 1250 would mean 99.50% minimum pure aluminium.

An example of the interpretation of the wrought aluminium designation system is as follows:

A typical, often used, alloy would be **6082**

    6 = group 6xxx (magnesium and silicon)

    0 = original alloy (not modified)

    82 = group specific allocator

Table 2.4 | Basic temper designation

Letter	Description	Meaning
F	As fabricated	Forming process with no special control over thermal or strain hardening
O	Annealed	Heat treated to give min. strength improving ductility and dimensionality
H	Strain hardened	Strengthened by cold working
W	Heat treated	Solution heat treated but produces an unstable temper
T	Heat treated	Thermally heat treated with or without additional strain hardening

Designations for cast aluminium alloys (ACs) differentiate from those used for WAs. Table 2.3 illustrates the current system.

ACs are not commonly used for load bearing structures and structural design of cast aluminium alloys is not covered by BS 8118 or BS EN 1999 (Eurocode 9). The structural use and structural suitability of ACs is therefore subject to resistance testing and laboratory-based determination of performance properties.

2.4.2   Temper designations of aluminium alloys

Further to the above classification an additional mark is used for the designation of aluminium alloys due to the availability of different tempers. All temper designations, specifications and properties can be found in the European/British Standard BS EN 515. Again the temper designation and treatment resulting properties are dependent for the two groups of HT or NHT alloys. The mechanical properties of HT alloys can be changed by heat treatment. Heat is therefore used to strengthen or soften the material. Softening is often applied to help with forming processes. If required, HT alloys can be re-heat-treated after forming to restore their original properties. Compared with HT alloys, the properties of NHT aluminium alloys can only be improved by cold-working after the last annealing operation. Strengthening by cold-working will be reduced by heat and cannot be restored after heating.

Five basic temper designations are used for the current aluminium alloy temper designation system. These groupings are represented by the letters F, O, H, W and T. A description of the basic temper designation system is given in Table 2.4. The groups for strain-hardened alloys (H) and thermally heat treated alloys (T) are further subdivided indicating the applied treatment or treatment combinations.

Subdivisions of the strain-hardened and HT aluminium alloys are done by adding numerical indicators to the preceding letters. The range of the strain-hardened alloys is H1–H4 and HX2–HX8. The sub-groups for the heat treated alloys are T1–T9. A summary and associated explanation is given in Table 2.5.

Table 2.5 | Temper designation system to current standards

Temper destination (Based on guidance given in BS EN 515:1993)

XXXX	-F	As fabricated	
	-O	Fully annealed (softened by heating)	
	-H1	Strain(Work)-hardened only	NHT
	-H2	Strain(Work)-hardened and partially annealed	NHT
	-H3	Strain(Work)-hardened and stabilised (by low temperature treatment)	NHT
	-H4	Strain(Work)-hardened and lacquered or painted	NHT
	-HX2	Quarter-hard	NHT
	-HX4	Half-hard	NHT
	-HX6	Three-quarter-hard	NHT
	-HX8	Fully-hard	NHT
	-T1	Cooled from an elevated termperature shaping process	HT
	-T2	Cooled from an elevated termperature shaping process, cold worked and naturally aged	HT
	-T3	Solution heat-treated, cold worked and naturally aged	HT
	-T4	Solution heat-treated and naturally aged	HT
	-T5	Cooled from an elevated termperature shaping process and artifically aged	HT
	-T6	Solution heat-treated and artificially aged	HT
	-T7	Solution heat-treated and over-aged	HT
	-T8	Solution heat-treated, cold worked and then artificially aged	HT
	-T9	Solution heat-treated, artificially aged and then cold worked	HT

HT=heat-treatable/NHT=non-heat-treatable

The commonly utilised aluminium alloy 6082-T6 from the earlier example to explain the wrought aluminium designation system is used to explain the adopted designation system:

6 = Group 6xxx (magnesium and silicon)

0 = original alloy (not modified)

82 = group specific allocator

T6 = heat treated and artificially aged

2.4.3 Aluminium alloy series

1xxx series, unalloyed aluminium, EN-AW-1xxx
Materials in this group are mostly pure aluminium with a purity of minimum 99% where the last two digits give the percentage of purity above 99%. Thus, the scope of purity ranges from 99.00–99.99%, with a reference number of 12**50** indicating a minimum purity of 99.50%.

The maximum possible tensile strength is about 150 N/mm². 1xxx series aluminium alloys are often used in plants when corrosion resistance is the dominant requirement. When rolled to foil thickness it is also used in the food and packaging industry.

2xxx series, copper alloys, EN-AW-2xxx
Alloys in this group are high-strength materials containing copper as the principal element, but also magnesium, manganese and silicon. In the T4 condition similar mechanical properties to mild steel can be achieved with a typical proof stress of 250 N/mm² and in the T6 condition a typical proof stress of 375 N/mm² giving a typical tensile strength of 460 N/mm². There are a few drawbacks to the good mechanical properties, such as: lesser corrosion resistance, reduced ductility, poor extrudability, unsuitable for arc welding and the higher price. 2xxx series aluminium alloys are mostly used for the aerospace industry.

3xxx series, manganese alloys, EN-AW-3xxx
With a tensile strength of approximately 200 N/mm², the 3xxx series alloys are not much stronger than pure aluminium. With a very high resistance to corrosion these alloys are used in the cladding of buildings (corrugated and profiled sheet material) and vehicle panelling.

4xxx series, silicon alloys, EN-AW-4xxx
The alloys of this series appear rarely in the form of structural elements or members. They are used for castings and weld filler wire with melting points lower than the parent material.

5xxx series, magnesium alloys, EN-AW-5xxx

This NHT alloy has a good combination of high strength and excellent resistance to corrosion. Its application is mainly in structural use and it has good weldability. Tensile strengths can exceed 300 N/mm^2. Typical products are sheet, plate and sheet-metal fabrications with uses for vessels, vehicles, ships and chemical plant.

6xxx series, magnesium–silicon alloys, EN-AW-6xxx

Containing magnesium and silicon these alloys have good all-round properties i.e. excellent extrudabtility, good resistance to corrosion combined with high strength. Tensile strength is in the region of 300 N/mm^2 with a proof stress of 250 N/mm^2. This material is weaker than mild steel and less ductile. This group includes the 6082 alloy which is widely used for building structures.

7xxx series, zinc alloys, EN-AW-7xxx

This series of zinc alloys display the highest strength of aluminium alloys. In the T6 condition it reaches a tensile strength of approximately $550–580 \text{ N/mm}^2$. Its mechanical properties are greatly improved compared to the 6xxx series alloys and HAZ softening at welds is less severe than for 6xxx series alloy. But again there are similar drawbacks to the 2xxx series such as: lesser resistance to corrosion, unsuitability for arc welding, an extrudability less than that of the 6xxx series and difficulty in fabrication. Thus, the 7xxx, which is mainly used in military or other specialised applications, requires a high degree of expertise to produce it and greater experience is needed to fabricate it.

2.4.4 Durability rating

Aluminium has a good resistance to corrosion in most environments and many chemical agents. This outstanding resistance to corrosion is based on its high affinity for oxygen. This attraction to oxygen causes the formation of a thin aluminium oxide film on its surface. This thin protective oxide film forms a shielding layer between the metal and its immediate environmental conditions. The physico-chemical stability of the oxide film determines the corrosion resistance of the aluminium. The permanence of this protective film is dependent on the pH value of the surrounding environment and it will be firm and protective within a pH range of about 4–8. In this environment the aluminium surface will maintain its original appearance and protection will not be required. Below pH 4 acid dissolution occurs and above pH 8 alkaline deterioration occurs, diminishing the aluminium oxide film and resulting in corrosion of the metal.

Where aluminium and aluminium alloys are exposed to more hostile environmental conditions protective precautions are necessary to ensure the designed and required lifespan.

For structural design BS 8118, and the new Eurocode 9 give guidance if and what protection of the material used is necessary and advised.

Aluminium and aluminium alloys with associated properties are listed in BS 8118 and Eurocode 9 with an indication of their durability ratings (see Table 2.6). The stated durability ratings are categorised into three durability ratings: A, B and C. These ratings are used to determine the grade of protection that is required. The required protection can be supplied as coats of paint, powder coating, or anodising.

Recommendations for protection according to exposure and material conditions are given in BS 8118 and BS EN 1999–1-1. Based on durability rating and exposure both standards show whether or not protection is required.

Table 2.6 | Durability rating of typical aluminium alloys used in structural design

Durability rating	Wrought aluminium alloys	Cast aluminium alloys
	EN AW 3004	EN AC 51300
	EN AW 3005	
	EN AW 3103	
	EN AW 5005	
A	EN AW 5049	
	EN AW 5052	
	EN AW 5083	
	EN AW 5454	
	EN AW 5754	
	EN AW 6005	EN AC 42100
	EN AW 6060	EN AC 42200
	EN AW 6061	EN AC 43000
B	EN AW 6063	EN AC 43300
	EN AW 6082	EN AC 44200
	EN AW 6106	
	EN AW 8011	
C	EN AW 7020	

2.4.5 Material selection

The choice of a suitable aluminium or aluminium alloy is the responsibility of the design engineer or structural engineer. Material selection is dependent on factors such as physical properties, strength, durability, formability and weldability. A further factor that might influence the selection of a particular aluminium alloy is the availability. The suitability of an aluminium alloy for structural design solution is based on a combination of detailed requirements such as: application, strength, shape and corrosion resistance. Comparison tables are available to assist the design engineer in selecting the most suitable aluminium alloy. Thus, material selection is a fundamental part of the structural design process.

A good knowledge and understanding of the material and its properties is important for the process of material selection. The elementary properties of available aluminium alloys have been illustrated in the previous sections of this chapter and should assist in the material selection procedure. A systematic approach should be taken to determining the most suitable aluminium alloy by comparing the design requirements against the properties of available, pre-selected or desired alloys.

For pre-selection purposes, the criteria listed below indicate when aluminium alloys are not suitable and should not be used:

- Service temperature exceeding 200°C or 100°C in combination with dominant loading applied.
- Required protection cannot be given.
- Is in high-grade acidic or base environments.
- Thermal or electric insulation is required or if thermal expansion must be minimised.
- Strength needs are > 500 kN/mm^2 or fatigue limits > 230 kN/mm^2.
- Low elastic deflection limits required.

2.5 Aluminium products

Aluminium and aluminium alloys are processed in various ways to produce aluminium products for industrial, commercial and private use. The manufacturing process for aluminium products includes more modern methods such as milling and extruding, but also the more traditional processes of forging and casting. Whilst forging and castings products are still widely used, especially for architectural and home or office use, milling products and extruded section are more often utilised for structural applications. The current versions of BS 8118 and Eurocode 9 reflect the preference for extruded and milled aluminium products and only give limited guidance for cast products.

Milled and extruded, but also drawn products used for structural application are subdivided into flat products, extruded products and tube products. Their main characteristics are the process of manufacture and heating used for the manufacture of specific products.

2.5.1 Flat products

Aluminium plate and sheet represent the group of flat products (see Figure 2.10). Plate and sheet products are manufactured in conventional rolling mills, similar to steel products, however, with much lower temperatures involved. In the milling process the aluminium ingot is first passed though a hot-rolling mill at a temperature of approximately 500°C. This process is repeated until the required thickness is achieved for the cold-rolling process. The end product is formed accurately by passing through the cold-rolling mill. Nevertheless, aluminium sheet and plates can be produced by hot-milling alone. Typical rolled products are plate and sheet, but also foil. Aluminium sheet products are used for panelling and roofing and have a typical thickness of 0.2–6 mm. Plate products are widely applied in structural uses within the construction industry, bridge building and automotive industry. Aluminium plate products have a typical thickness from 4 mm upwards.

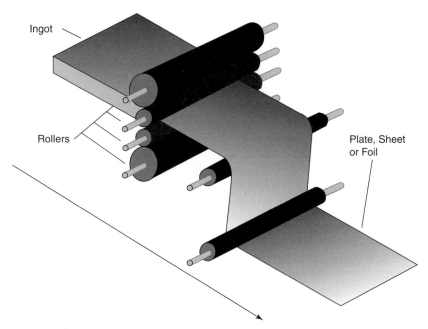

Figure 2.10 | Manufacture of flat products (Reproduced with permission of IAI, London, UK)

Sheet and plate products, their condition, material properties and tolerances are specified in BS EN 485, Parts 1–4 Aluminium and aluminium alloys – Sheet, strip and plate:

- Part 1: Technical conditions for inspection and delivery.
- Part 2: Mechanical properties.
- Part 3: Tolerances on dimensions and form for hot-rolled products.
- Part 4: Tolerances on shape and dimensions for cold-rolled products.

2.5.2 Extruded products

Aluminium extrusions are produced by an aluminium billet being forced through a die with an opening shaped to give the desired profile section. The billets are heated to near 500°C and are then rammed with a high pressure of 500–700 MPa to extrude through the shaped opening in the die (see Figure 2.11).

The extrusion process allows aluminium sections to be formed in an almost unlimited range of shapes. Typical sizes for aluminium extrusions are in the range 10–800 mm with a cross-sectional shape designed to suit the intended application and a minimum thickness of about 5 mm. Typical shapes are solid and hollow sections with many standard and stock sizes and shapes. I, C and Z sections are also widely available from a variety of stockists. Extruded sections have the benefit that they allow the designer to specify the section required for the task and parts of the section can easily be strengthened to suit requirements. For example, functional features can be

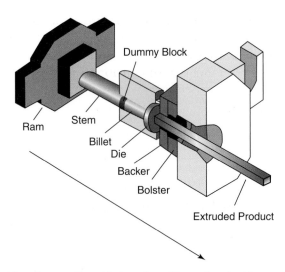

Figure 2.11 | Manufacture of extruded products (Reproduced with permission of IAI, London, UK)

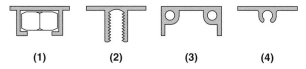

Figure 2.12 | Functionality added to extrusion shapes

added to combine structural strength with allowances for fabrication, assembly and erection. Typical examples of added functionality are bolt or nut slots (1), screw slots (2) and screw ports (3) and (4), as shown in Figure 2.12.

Extrusion speed and applied pressure must be carefully controlled to prevent unnecessary stresses and exit temperature in the extrusion. Many extruded section length require post-treatment and correction to remove manufacture stresses and distortions. Heat treatment can be done by spraying sections with water as they come forward from the extrusion process. However, if a more accurate temperature control is required, extruded lengths have to be cut, re-heated and quenched in a tank.

Extrusion products, their condition, material properties and tolerances are specified in BS EN 755, Parts 1–9 Aluminium and aluminium alloys – Extruded rod/bar, tube and profiles, as follows:

- Part 1: Technical conditions for inspection and delivery.
- Part 2: Mechanical properties.
- Part 3: Round bars, tolerances on dimensions and form.
- Part 4: Square bars, tolerances on dimensions and form.
- Part 5: Rectangular bars, tolerances on dimensions and form.
- Part 6: Hexagonal bars, tolerances on dimensions and form.
- Part 7: Seamless tubes, tolerances on dimensions and form.
- Part 8: Porthole tubes, tolerances on dimensions and form.
- Part 9: Profiles, tolerances on dimensions and form.

2.5.3 Tube products

Aluminium tube sections include hollow section with uniform tube wall thickness shaped round, oval, hexagonal, square and rectangular. Extruded tubes are produced using a bridge or porthole die. Due to heat and applied pressure the metal flows around the bridge and is joined again by a hot pressure welding chamber. This has the disadvantage of not being strictly seamless and prevents the use of these tubes for some structural application.

Figure 2.13 shows a simplified, typical bridge die and its associated parts. Bridge dies for tube extrusions are designed to suit the desired tube shape with an infinite number of possible shapes and sizes. It is

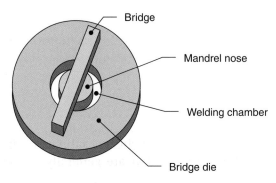

Figure 2.13 | 'Bridge die' for tube extrusion

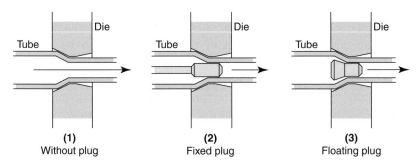

Figure 2.14 | Tube drawing process

also possible to have several extrusion shapes within one die if the profiles are extruded in parallel.

For uses requiring seamless and high quality tubes, drawn tubes are necessary to fulfil these requirements. The cold-drawn manufacture method starts off with an extruded tube reduced by the drawing process. Following the extrusion process, tubes are cold drawn using dies and plugs to the required shape, size, tolerances and mechanical strength. The final tube is of an entirely homogeneous metallurgical structure. This type of cold-working enables application to HT and NHT materials.

The drawing method usually consists from using a fixed plug (see Figure 2.14). The outer diameter (OD) of the drawn tube is then defined by the die and the inner diameter (ID) by the plug installed on a long rod passing through the original tube. A more modern method has been developed using a floating plug. This method enables long lengths to be drawn at more efficient speed with the tube feed onto a coil.

The cheapest way to manufacture thin tube is by tube welding, where a cold-reduced strip is rolled to a circular shape and seam welded by an automated welding machine.

2.6 Element fabrication

Aluminium goods are normally supplied as semi-finished products, mostly as sheet, plate, extrusions and tube products as discussed in Section 2.5. All of these products are then further shaped and fabricated to produce the required shapes or elements using a wide range of fabrication processes. Generally, all fabrication methods used in steel fabrication can be applied to aluminium goods. However, the softer aluminium allows faster and cheaper fabrication if compared to steel fabrication. The processes that are generally employed are cutting, sawing, drilling, punching, bending, machining and welding.

2.6.1 Sawing and cutting

Aluminium products can very easily be sawn. The most common types of sawing are using a circular saw or band saw. For both methods the saw blades should be carbon tipped and must be lubricated by spraying a stick wax onto the blades during sawing.

For more precise shaping laser, plasma gas and water jet cutting is used to achieve high precision cuts. Laser and plasma gas cutting required further treatment and cleaning of the cutting edge to remove contaminations and dross. The advantage of water jet cutting is the lack of heat and therefore no alteration to the properties of the aluminium.

A further option is shear cutting on a guillotine, as used for straight lengthening of plate and also flat sheet products. Shearing produces a clean and accurate cutting edge.

2.6.2 Drilling and punching

Holes and slots can be made in aluminium products by drilling and punching. Whilst standard steel-drill can be used for drilling aluminium it is advisable to use aluminium-suitable drills for better chip removal. Drilling is mostly carried out with drilling fluid similar to sawing: fluid stick wax is sprayed onto the drills during drilling. It is common practice to drill holes in aluminium with a smaller drill than required and ream to the exact diameter that is required for the hole. The reason for this procedure is that holes drilled in the softer aluminium can become oversized by up to 0.5 mm whereas holes drilled in steel are usually accurate to drill size.

Punching is a fast, effective and clean method to produce holes in all shapes and slots in aluminium products. Similar to shear cutting, punching produces a clean and accurate opening. Punching does not necessarily require lubrication. However, lubricated punching produces a better result that punching without lubricant.

2.6.3 Machining and forming

The physical and mechanical properties of aluminium alloys allow machining of most aluminium products. Density, high thermal conductivity, modulus of elasticity and low shear modulus are beneficial to the machining process. Conversely, soft alloys can cause chipping problems such as build-up at the machining tool or blockage. Aluminium alloys that are especially suitable for machining are known as free machining alloys. They are alloys to which certain elements have been added that help to break up the machining chips, allow lower power and produce better finishes and longer tool life. Machinability in aluminium alloys is rated on a scale A–E (devised by the AA Inc.) which is related to the size of the machining chips formed and the resulting machined surface finish. Heat treatment can greatly improve machinability. Typical alloys suitable for machining are 2014A, 6082 and 7075. Lubrication is an important factor in machining aluminium products. Aside from the primary function of lubricating, lubricating fluid is also used to prevent swarf from bonding to tools and to remove swarf from the point of machining.

Forming is another fabrication process that can be used, mostly with sheet and plate products. Forming can be defined as plastic deformation of a mass of material to alter form and/or properties without changing mass and material. The two main methods of forming are: bending and folding as indicated in Figure 2.15. Whilst folding consists of several bending actions to produce the folded part, bending can often be executed in one operation. Bending and folding processes are often irreversible processes with part of the energy applied for the forming process converted into heat which causes the aluminium product to warm up.

The problems of cracking and resilience during the bending of aluminium sheet and flat products need to be taken into account. Resilience can be described as the angle of reverse movement after removal of the bending force and is directly dependent on the elastic-plastic forming behaviour of the material used. The main

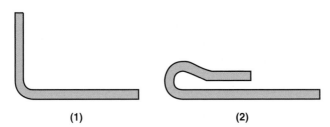

(1)   (2)

Figure 2.15 | Typical forming processes

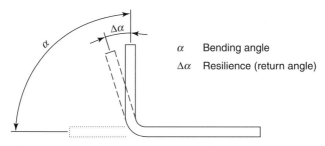

Figure 2.16 | Resilience to bending/springback

parameters that control the amount of resilience are: the modulus of elasticity, yield strength, the strain hardening coefficient and the ratio of the bending radius to the thickness of the material (sheet) (see Figure 2.16).

Minimal resilience (spring-back) can be achieved by:

* increasing the modulus of elasticity
* decreasing the yield strength
* limiting the ratio of bending radius to sheet thickness

Resilience after bending of sheet or bar material can be balanced by over-bending as shown in Figure 2.17.

Bending cracking can occur on the outside face of the bend sheet. Subject to tensile force, the cracking of the outside face of the aluminium sheet is the result of exceeding the allowable tensile stress within the sheet. The bendability of aluminium sheet and plate products depends on the type of alloy, temper, material thickness and bending

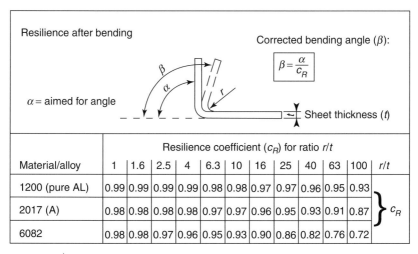

Material/alloy	1	1.6	2.5	4	6.3	10	16	25	40	63	100	r/t
1200 (pure AL)	0.99	0.99	0.99	0.99	0.98	0.98	0.97	0.97	0.96	0.95	0.93	
2017 (A)	0.98	0.98	0.98	0.98	0.97	0.97	0.96	0.95	0.93	0.91	0.87	c_R
6082	0.98	0.98	0.97	0.96	0.95	0.93	0.90	0.86	0.82	0.76	0.72	

Figure 2.17 | Tooling angle to bending angle

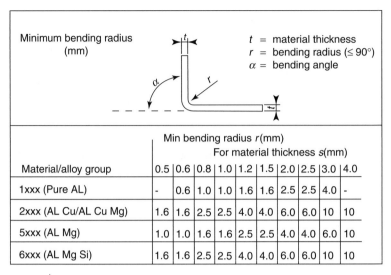

Material/alloy group	Min bending radius r(mm) For material thickness s(mm)									
	0.5	0.6	0.8	1.0	1.2	1.5	2.0	2.5	3.0	4.0
1xxx (Pure AL)	-	0.6	1.0	1.0	1.6	1.6	2.5	2.5	4.0	-
2xxx (AL Cu/AL Cu Mg)	1.6	1.6	2.5	2.5	4.0	4.0	6.0	6.0	10	10
5xxx (AL Mg)	1.0	1.0	1.6	1.6	2.5	2.5	4.0	4.0	6.0	10
6xxx (AL Mg Si)	1.6	1.6	2.5	2.5	4.0	4.0	6.0	6.0	10	10

Figure 2.18 | Approximate (minimum) bending radius

radius. The minimum allowable bending radius is a direct function of the alloy, its temper and its material thickness. It is therefore important to determine the minimum allowable bending radius for the specific aluminium product and material used. Typical minimum allowable bending radii for a bending angle of 90° are illustrated in Figure 2.18.

BS EN 485–2:2004 requires aluminium sheet, strip and plates to be capable of being bent cold through an angle of 90° or 180° around a pin having a radius of k times the material thickness t without cracking. Recommended values of the minimum bend radii for different alloys, tempers and thickness are given in Tables 2–39 of BS EN 485–2:2004 (see Figure 2.18).

2.6.4 Jointing

The jointing methods usually used in aluminium construction are: bolting, riveting, welding and adhesive bonding (gluing). However, these jointing types are not exhaustive. Depending on the type of structure, loading, material used and required stiffness, the designer can chose from a wide range of jointing methods. In general, the jointing can be subdivided into: mechanical joints, welded joints and bonded joints.

2.6.4.1 Mechanical joints

Bolting, screwing, riveting, interlocking, clinching and folding are used for mechanically joining aluminium elements. Whilst each industry applies its own fixing methods, bolting, screwing and riveting are the most commonly used in the construction industry. Special care and attention is required when choosing the fastener material as bi-metallic corrosion could weaken or destroy the joint over time. For bolted connections aluminium or stainless steel bolts are recommended in aluminium construction. Stainless steel bolts are the preferred material and are available in austenitic stainless steel, which presents the best solution. But the use of stainless steel bolts in A2 and A4 material grade is also permitted. Galvanised steel bolts can only be utilised indoors or in dry environments. However, it is advisable to privilege stainless steel bolts over galvanised steel bolts. The extra cost of the stainless steel bolts will be relatively small when compared to the overall costs of the structure. Other than with a steelwork bolted connection, the aluminium structure bolted joint should have washers provided at head side and thread side i.e. nut side.

Riveting is another frequently used method of jointing in aluminium construction. When joining thin gauge sheet and materials with thin wall thickness, riveting is preferred to welding because of the strength loss due to HAZs. Aluminium rivets are used in reamed holes and are cold-driven and closed. Riveting should not be used to transfer tensile forced via the rivets as aluminium rivets have only limited tensile strength and are best employed in shear conditions. Aside from the conventional rivets in use, special rivets for joining flat elements and especially sheet are available, such as pop rivets (see Figure 2.19) and Chobert rivets (see Figure 2.20). The application of pop rivets and

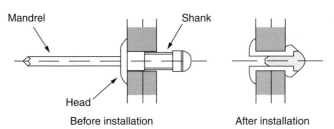

Figure 2.19 | Pop-riveting

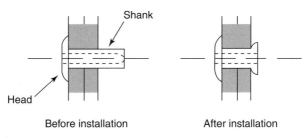

Figure 2.20 | Chobert riveting

Chobert rivets is known as blind riveting as all of the riveting can only be done from one side.

2.6.4.2 Welded joints

The most common way of joining aluminium is welding. Most alloys of aluminium can easily be welded. However, welding requires careful planning and preparation of the materials. As aluminium forms a tough oxide layer on its surface, and this protective layer has a higher melting point than aluminium, it must be removed before welding. It is removed using chemical, mechanical or electrical methods and reforming must be prevented before welding can be completed. Due to the high thermal conductivity of aluminium, heat needs to be applied at a rate four times that needed for steel. Aluminium has a linear expansion coefficient twice that for steel, which must be considered when welding material that has been restrained and cannot expand freely. Aluminium has a relatively low melting point and unlike steel, it does not change colour as its melting point is approached. Consequently, care must be taken not to overheat or melt aluminium during welding. However, welding reduces the mechanical properties of aluminium in the HAZ. As a rough guideline, this area extends approximately 25 mm from the weld.

For economic and quality reasons, metal inert gas (MIG) and tungsten inert gas (TIG) welding are the recommended methods for welding aluminium. Further guidance can be found in Section 2.8 of this chapter.

2.6.4.3 Bonded joints

As welding reduces the properties in the HAZ, aluminium parts are being increasingly joined with adhesives. Adhesives are now commonly used for joining aluminium in structural applications. New technologies in the area of synthetic adhesives promise to further increase the importance of adhesive bonding for aluminium. When adhesive bonding is used with aluminium, it is generally found that no bonding occurs between the adhesive and the aluminium metal. Rather, the adhesive bonds to the aluminium oxide layer. Acid etching can be used as a surface preparation to create a bond directly to the aluminium. Surface preparation is dependent upon the type of adhesive being used and should be done in accordance with the manufacturer's recommendations. Also, it is not adequate to simply use adhesive on a joint that would otherwise be welded or mechanically fastened. Joint design for adhesive bonding should allow for maximum surface contact between the adhesive and the aluminium. Adhesive bonding works best when the forces are predominantly pure shear, tension or compression. The use of lap joints is common as they have a large joint surface area, load predominantly in tension and avoid cleaving or peeling forces. Guidance in regards to adhesive bonding is given within the British Standard 8118 and Eurocode 9.

2.7   Heat-affected zone softening

Most aluminium alloys used in structural applications have mechanical properties based or improved due to cold-working or heat treatment. When welding these aluminium alloy members, heat generated by the welding process reduces material properties in the vicinity of the welds. The zones affected by this softening are therefore called heat-affected softening zones.

Both British Standard 8118 and Eurocode 9 give detailed guidance for structural designers how to make allowances for the loss of strength within HAZs. The reduction in strength can be severe with a local drop in the parent material's strength by approximately one half. Only for parent material supplied in the annealed or T4 condition can HAZ zone softening be ignored. For structural design the extent of heat-affected softening and its severity must be known.

The severity of heat-affected softening is expressed by application of a reduction factor, where strength properties of the parent material are reduced by the earlier mentioned reduction factor. A typical reduction factor for 6082 aluminium alloy in T6 condition taken from BS 8118: Part 1: Table 4.5 would be 0.5. Thus a strength loss of 50% must be considered to allow for welding.

It is also important for structural design to determine the extent of the heat-affected softening zones. A rule of thumb can be applied to identify the affected zone, better known as the 'one-inch rule'. It is assumed that the heat-affected softening zone extends one inch from the centre of the weld giving a 'z' of approximately 25 mm. If the effect of HAZ on the resistance of the section is relatively small, the sole use of the one-inch rule is acceptable. However, where simple calculations based on the one-inch rule show significant softening, a more accurate approach is needed for final design checks.

The science of heat-affected softening is not clearly defined and various methods are available to estimate the extent of the affected areas (see Figure 2.21). In some cases it can be beneficial to produce a prototype and determine the extent of the heat-affected softening by experimental methods and hardness surveys.

2.8   Corrosion

Aluminium has a good resistance to corrosion in most environments and many chemical agents. This outstanding resistance to corrosion is

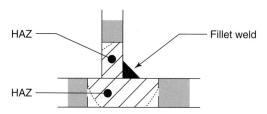

Figure 2.21 | Extent of HAZ

based on its high affinity for oxygen. This high affinity of aluminium for oxygen results in the development of an alumina film, ensuring that new films will develop in damaged regions and areas, meaning an ideal situation arises where the metal is protected by a very thin, self-repairing protective layer. However, even with the earlier mentioned good corrosion resistance and durability, corrosion is still a possibility. Damage to the surface and protective layer can occur due to weathering, material imperfections and bi-metallic reaction with other materials in contact with the aluminium alloy elements.

2.8.1 Pitting corrosion

Pitting is the most common form of corrosion of aluminium and aluminium alloys. Pitting is a localised type of corrosion and occurs at defects within the surface protective layer and where the film is damaged mechanically, such as scratches. If self-repair of the protective layer is not possible, due to lack of oxygen, pitting initiates. Pitting corrosion results in small, often microscopic, holes and cavities in the material with total penetration possible. The pitting corrosion process can be divided in two separate actions, the *initiation* and the *propagation*. The corrosion rate decreases over time, with the corrosion products suffocating the process.

Due to the small size of the pits that are created, pitting corrosion is often not noticed early on. However, white corrosion product, $Al(OH)_3$, deposited at the pit perimeter is a good indicator of pitting corrosion. This white powder may cover the pits entirely, appearing as pale dots at the material surface.

2.8.2 Crevice corrosion

Crevice corrosion is a localised form of corrosive attack similar to pitting corrosion. Crevices, that allow crevice corrosion, can form at narrow openings, spaces between metal surfaces or spaces between metal surfaces and non-metal surfaces. Examples of crevices are: flanges, washers, threaded joints and gaped joints but also cracks and seams. Differential aeration is an important factor in the mechanism of crevice corrosion leading to the formation of acid within the crevice.

Because of the often hidden initiation of crevice corrosion behind material layers, gaskets, seams or cracks it is difficult to detect. Nevertheless, crevice corrosion should be addressed as localised failure of the material at joints, laps etc. are the consequences.

2.8.3 Intergranular corrosion

Intergranular corrosion is defined as a localised corrosion attack in which a path is corroded along the grain boundaries of the aluminium or aluminium alloy, while the majority of the grains remain unaffected. This form of corrosion is usually associated with chemical segregation effects or specific phases precipitated on the grain boundaries. Reheating a welded component for stress relieving is a common cause of this problem. Intergranular corrosion is also often associated with high-strength aluminium alloys. But alloys that have been extruded or otherwise worked heavily are highly susceptible to this type of corrosion. Intergranular corrosion can often be found in alloys belonging to the groups. 2xxx, 5xxx and 7xxx.

2.8.4 Exfoliation corrosion

Exfoliation corrosion is a particular form of intergranular corrosion associated with high-strength aluminium alloys. Generation of corrosion particles forces the layers apart and causes the metal to intumesce. Metal flakes can be pushed up and peel (exfoliate) from the surface. Exfoliation corrosion is most common in the HT Al–Mg–Cu and Al–Zn–Mg–Cu alloys.

2.8.5 Filiform corrosion

Filiform corrosion usually appears underneath surface layers such as paint or directly on the surface. It appears as random non-branching white tunnels of corrosion product and it is a structurally insensitive form of corrosion. As with the earlier types of corrosion, filiform corrosion initiates at defects (scratches) to surfaces and coating. Filiform corrosion spreads by undermining i.e. the aluminium or aluminium alloy surface just below the protective layer or coating is corroded away. Filiform corrosion depends on the relative moisture of the air and the quality of the surface treatment preparation prior to coating.

2.8.6 Stress corrosion (stress corrosion cracking)

Stress corrosion cracking (SCC) is the failure by cracking caused by intergranular or transgranular corrosion and applied tensile action to a structural member. SCC is a structurally sensitive form of corrosion as failure can take place suddenly without warning signs. Due to the granular corrosion attack failure will happen well below normal fracture stresses. SCC can be summarised to an alloy being sensitive to SCC, an aggressive environment and sufficient tension applied.

2.8.7   Poultice corrosion

Poultice corrosion is caused if aluminium or aluminium alloy materials are brought into contact with wet or damp matter where the natural protective layer and also the metal underneath the protective layer is corroded. Poultices can be formed by organic or inorganic substances containing high moisture content, subsequently creating the environment for corrosion to thrive.

A special form of poultice corrosion can be initiated by contact of aluminium or aluminium alloys with wood, concrete and masonry in damp settings. All three materials might contain chemicals for treatment or normal manufacture and workings, which can accelerate the corrosion attack. Contact with these materials must be avoided if no suitable protection is applied.

Poultice corrosion is not a distinct form of corrosion, rather it is a special case of crevice corrosion. But it is important to understand this type of corrosion and the possible causes. It is also a type of corrosion that can easily be prevented by good design and regular maintenance.

2.8.8   Galvanic corrosion

Galvanic corrosion, also known as bi-metallic corrosion, occurs when metallic materials are in contact in the presence of a corrosive electrolyte, corrosion will increase on the least noble material (the anode) and decrease on the noblest (the cathode). Aluminium is generally a base material i.e. less noble than most of the other metallic materials used in construction. Exceptions are beryllium, magnesium and zinc. Thus, when aluminium is in contact with most other metallic materials, and with moisture present, accelerated corrosion will occur. The rate of corrosion and severity of corrosion is dependent on the properties of the more noble material. When in contact with steel and iron, aluminium will undergo severe corrosion. However, copper metals and copper-based metals such as brass and bronze will cause a more relentless corrosion compared to steel-based metals.

Good precautionary practice is vital to protect aluminium elements from this very serious corrosion attack. Without precautions i.e. isolation of the two metals connected, austenitic stainless steel can be used safely.

Aluminium in contact with magnesium and zinc would lead to corrosion of the less noble metal, in this cases magnesium and zinc. This will cause corrosion problems with the galvanised coatings that are often used to protect bolts and screws.

2.8.9   Corrosion prevention

Simple measures to prevent corrosion of aluminium and aluminium alloy elements can be considered and applied. Some of the common preventative methods will be considered in Chapter 9. Some considerations to aid with corrosion prevention are listed below (see also Table 2.7):

- Select resistant materials.
- Protect materials by surface treatment.
- Protect material prior to fabrication and transport.
- Use suitable fabrication methods.
- Avoid stress concentration.
- Avoid bi-metallic contact.
- Prevent moist and damp conditions.
- Prevent contact with wet and damp substances/matter.
- Keep structures clean and dry.
- Allow for drainage if exposed to rain or spray.
- Allow for access for maintenance and repair.
- Ask the experts.

Table 2.7 | Electro-chemical series of metals used for construction purposes

Most noble	Gold
Cathodic	Platinum
	Silver
	Nickel
	Copper
	Brasses
	Lead
	Tin
	Cast iron, high strength steel
	Mild steel
	Cadmium
	Titanium
	Aluminium
	Beryllium
	Zinc
Anodic	Magnesium
Less noble	Lithium

3

STRUCTURAL ANALYSIS

Introduction

Structural analysis is the systematic study of the interaction of structural materials, structural elements, their structural properties and physical conditions such as supports, dimensions and loading. By applying the physical laws, the behaviour of a structure can be calculated and properties changed if necessary. The results of this procedure are then used to design structural elements such as: plates, beams and columns to comply with the recommendations given in material-specific design codes. The design codes for structural aluminium are BS 8118 and BS EN 1999, namely Eurocode 9.

In general, structural design of structures is ruled by two main design approaches: the allowable stress design and the limit state design (LSD). Whilst the allowable stress design uses working loads to determine the actual stresses in a member, the LSD applies load factors to the working loads and determines the design resistance of a member.

Irrespective of the structural design route taken, the structure must be designed to perform satisfactorily during its lifetime i.e. withstand all applied loading and not deform excessively. Compliance with the design codes should generate a satisfactory structure. With the aid of modern computerised analysis methods, more rigorous approaches are possible, that find the potential overload before failure. Eurocode 9 only gives a little guidance on plastic analysis, and these methods should be considered with caution as miscalculation can have serious consequences.

3.1 Limit state design

LSD originated in the former Soviet Union and was first introduced in Europe in the 1960s. Until recently most of the design codes utilised the permissible stress design method (often termed elastic design) to check whether or not a structure or part of the structure assessed would be fit and safe to use. With the introduction of the modern design codes, some of which are the Eurocodes, the design methods are now based on the LSD method.

The approach used in LSD is for the engineer to analyse, check and design the structural elements to satisfy and comply with preset conditions as specified in the design codes. The LSD consists of two limit states: the ultimate limit state (ULS) for structural strength, and the serviceability limit state (SLS) for serviceability i.e. deformation and vibration. Both design codes, BS 8118 and BS EN 1999 require structural elements to be analysed and designed first, not to exceed the deflection limits set in SLS, and secondly for the applied and factored loadings in ULS to be less than the determined element design resistance. Table 3.1 summarises the basic requirements for ULS and SLS criteria.

BS EN 1990 'Basis of structural design', often referred to as Eurocode 0, further subcategorises the ULS into following limit states:

- EQU: loss of static equilibrium of the structure
- STR: internal failure or excessive deformation of the structure or a structural member
- GEO: failure due to excessive deformation of the ground
- FAT: fatigue failure of the structure or a structural member

To determine whether or not a structure or components of a structure is/are performing satisfactory i.e. they are safe to use, the above-mentioned states and associated actions on the structure or individual structural member must be analysed. Applied member actions are then compared to the structural material properties giving the member the capacity to resist applied actions. A satisfactory design will demonstrate a structural capacity greater than the applied design actions.

3.2 Load factors and loadings

Structures must be designed to be able to carry all applied loads (actions) and not deform excessively. Thus, the determination of the loads acting on a structure is an important element of structural

Table 3.1 | Basic criteria for limit state design

State	BS 8118	BS EN 1990
ULS	Static strength	Static strength
		Fatigue
SLS	Deformation	Deformation
	Fatigue	Durability
	Vibration	Vibration

design. Computation of the maximal loads and combined loads can be complex and time consuming, as loads are dependent on the type of structure, its location and often also the method of erection.

Guidance to establish the basic loads acting on a structure and their combination is given by BS 648 and BS 6399 and Eurocode 0 and Eurocode 1. Whilst calculation procedures to define design and combined loading action on structures are differently specified in British Standards compared to those given in the Eurocodes, both design norms use the same basic load types i.e. types of actions. These types of actions are divided into: permanent actions, variable actions and accidental actions as illustrated in Table 3.2.

Table 3.2 | Classification of actions to BS EN 1990, Eurocode 0

Action	**Symbol**	
Permanent	G/g	Self-weight of structures
		Fixed equipment
		Road surfacing
		Shrinkage and uneven settlement
Variable	Q/q	Imposed loads
		Wind actions
		Snow loads
		Traffic loads
Accidental	A/a	Explosion
		Impact from vehicles
		Seismic actions

G, Q, A = concentrated load

g, q, a = uniformly distributed load

3.2.1 Loading

The main types of loading used for structural design are usually permanent and variable loadings. These two types of loadings are also called static loads as they tend not to change rapidly during a specified period of time. The extents of these loads are either given in the appropriate sections of design codes and standards or must be calculated via a set of equations provided in the relevant codes and standards.

The most common loads required for design of structures are:

* self-weight of the structural elements (beams, columns, rafters, slabs etc.)
* weight of the fixtures and installations (decking, insulation, services, finishes)

- imposed loads
- wind loads
- snow loads
- maintenance loads (access of workmen and equipment to roofs, ceilings)

The vast majority of structures can be designed safely by applying the above loads and their combinations. More specialised structures such as: bridges, tunnels, high-rise towers, car parks and other highway structures, to name but a few of them, will require further loading to be applied which falls into the category of dynamic loads i.e. accidental loads and seismic loads.

Table 3.3 provides a summary of where to find specific loads and guidance on how to calculate the required loading values:

Permanent actions (often referred to as dead load in British Standards): There is very little difference in the determination of permanent loads to British Standards and the new Eurocodes, especially for self-weights of materials. These loads are simply taken from the relevant sections of the design codes (shown in Table 3.3), where it is considered adequate to use weights based on their mean/bulk densities. Some typical mean densities are listed in Table 3.4.

Variable action (live load or imposed load in British Standards):
Whilst the current British Standard BS 6399 gives a specified value for the imposed load to be applied to the structure or structural element, the new Eurocode calculates a set of four representative values for variable actions.

Table 3.3 | Standards/sources for loadings

	British Standard	**Eurocode**
Self-weight of materials	BS 648	BS EN 1991–1–1
Imposed loads	BS 6399: Part 1	BS EN 1991–1–1
Wind loads	BS 6399: Part 2	BS EN 1991–1–4
Snow loads	BS 6399: Part 3	BS EN 1991–1–3

Table 3.4 | Mean densities for construction materials

Material	**BS 648**	**BS EN 1991–1–1**
Aluminium	2771 kg/m^3	27 kN/m^3
Concrete	2307 kg/m^3	24 kN/m^3
Steel	7849 kg/m^3	78.5 kN/m^3
Timber (softwood)	480 kg/m^3	4 kN/m^3

The principal representative value is the characteristic value (Q_k, q_k). This can be determined statistically or a nominal value can be used. The other representative values are the combination, frequent and quasi-permanent values, calculated by applying the factors ψ_0, ψ_1 and ψ_2, respectively. Hence, the values are defined as specified in BS EN 1990 for the combination value $(\psi_0 Q_k)$ for the frequent value $(\psi_1 Q_k)$ and for the quasi-permanent value $(\psi_2 Q_k)$. The combination value $(\psi_0 Q_k)$ of an action takes account of the reduced probability of the simultaneous occurrence of two or more variable actions. The frequent value $(\psi_1 Q_k)$ is such that is should be exceeded only for a short period of time and is used primarily for the SLSs but also for the accidental ULSs. The quasi-permanent value $(\psi_2 Q_k)$ can be exceeded for a longer period of time. Alternatively, it may be considered as an average loading over time and is used for the long-term effects at SLSs and accidental/seismic ULSs.

The values of the ψ-factors for the UK are taken from the UK National Annex to Eurocode 0 as shown in Table 3.5.

Table 3.5 | ψ-factors to Eurocode 0, National Annex to BS EN 1990

Action	ψ_0	ψ_1	ψ_2
Category A: domestic and residential areas	0.7	0.5	0.3
Category B: office areas	0.7	0.5	0.3
Category C: congregation areas	0.7	0.7	0.6
Category D: shopping areas	0.7	0.7	0.6
Category E: storage areas	1.0	0.9	0.8
Category F: traffic area, vehicles ≤ 30 kN	0.7	0.7	0.6
Category G: traffic area, vehicles ≤ 160 kN	0.7	0.5	0.3
Category H: roofs	0.7	0	0
Snow loads: sites located at $H > 1000$ m	0.7	0.5	0.2
Snow loads: sites located at $H \leq 1000$ m	0.5	0.2	0
Wind loads on buildings	0.5	0.2	0
Temperature (non-fire) in buildings	0.6	0.5	0

Example 3.1 Variable actions/imposed loading to Eurocode 1 and BS 6399:
Criteria: characteristic imposed floor loading for domestic property
(1a)   Imposed load to BS 6399: Part 1 (*use BS 6399: Part 1, read from Table 1, Category A*):

uniformity distributed load	$IL_{UDL} = 1.5$ kN/m²
concentrated point load	$IL_{PL} = 1.4$ kN

(1b)   Imposed load/variable action to Eurocode 1 (BS EN 1991–1-1) (*use National Annex to BS EN 1991–1–1; obtain category from Table NA.2 = A.1; read from Table NA.3, Category A.1*):

uniformly distributed load $q_k = 1.5$ kN/m²
concentrated point load $Q_k = 2.0$ kN/m²

3.2.2   Load factors

Load factors are applied to the characteristic loads or actions to compute the design values of the actions to be used in structural design. In principle, this procedure is the same for designs complying with British Standards or Eurocodes. Each standard uses different values for the factors to be applied to find the design loads for the SLSs and ULSs.

When designing aluminium structures to British Standards, the relevant load factors are specified in BS 8118: Part 1: Clause 3.2.3 *Factored loading*. According to Clause 3.2.3 the overall load factor γ_f is calculated as follows:

$$\gamma_f = \gamma_{f1} \times \gamma_{f2} \tag{3.1}$$

Where γ_{f1} and γ_{f2} are partial load factors and their values can be found in Tables 3.1 and 3.2 of BS 8118. For standard design situations with the imposed load or wind action giving the most severe loading action on the structure or component, the typical overall load factors can be summarised as in Table 3.6.

Table 3.6 | Overall load factors to BS 8118

	SLS	**ULS**
Dead load	1.0	1.2
Imposed load	1.0	1.33
Wind load	1.0	1.2

Table 3.7 | Load factors to Eurocode 0, BS EN 1990

	SLS	**ULS**
Permanent action(s)	$1.00\ G_k\ (g_k)$	$1.35\ G_k\ (g_k)$
Variable action (leading)	$1.00\ Q_k\ (q_k)$	$1.50\ Q_k\ (q_k)$
Variable action(s) (others)	$\psi_{0/1/2} \times Q_k\ (q_k)$	$\psi_{0/1/2} \times 1.5 \times Q_k\ (q_k)$

In contrast to BS 8118, the load factors for designing aluminium structures are given in the Eurocode 0, BS EN 1990 and its National Annex. The main difference between the overall load factors from BS 8118 and the load factors to be used in designs to Eurocode 9 is the allowable reduction of load factors for non-leading variable actions. Thus, only the major i.e. leading variable action is fully factored whilst the other variable actions are reduced by the application of the appropriate values of $\psi_{0/1/2}$. Typical load factors for standard designs are shown in Table 3.7.

Example 3.2: ULS design actions/factored loading to Eurocode 0 and BS 8118:

Criteria: structural facade for office property, aluminium rectangular tube column supporting floor and wall panel above but also resisting wind pressure action on glazing fitted between the aluminium columns.

Characteristic loading/actions: (on one column)

> permanent/dead load = 7.50 kN axial
> variable/imposed load =10.0 kN axial
> variable/wind load = 2.00 kN/m uniformly distributed

(2a)   Factored loading to BS 8118:
overall load factors (ULS):

> dead load = 1.2, imposed load = 1.33, wind load = 1.2

design loading (ULS):

> dead load = 7.50 × 1.2 = 9.00 kN
> imposed load = 10.0 × 1.33 = 13.3 kN
> wind load = 2.00 × 1.2 = 2.4 kN/m

(2b)   Design actions to Eurocode 0:
load factors (ULS): permanent = 1.35, leading variable = 1.5

> Other variable $\psi_0 \times 1.5$

design actions (ULS):

> permanent actions = 7.50 × 1.35 = 10.125 kN
> variable actions (leading) = 10.0 × 1.5 = 15.000 kN
> variable actions (others) = 2.00 × 1.5 × 0.7 = 2.100 kN/m

As seen in Example 3.2, the design loads generated with the procedure of Eurocode 0 generates higher values for the design actions for the ULSs. Figure 3.1 compares load factors for typical design combinations such as dead load (permanent actions) combined with imposed load (variable actions) but also dead load combined with imposed load and wind load. The appropriate factors have been

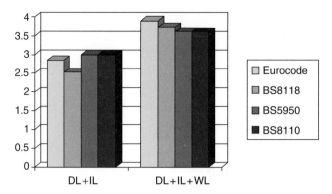

Figure 3.1 | Load factor comparison index, ULS

added together to result in a factor comparison index for each design standard.

3.3 Member properties

Structures are made of three distinct components: members, connections and supports. The primary load-bearing components are the members, usually prismatic elements made of a single material, for example aluminium. As discussed in Chapter 2 it is a fundamental requirement to know the structural properties of material and members used in the structural design. The most important material structural properties are:

- Modulus of elasticity (E) is also known as the elastic modulus or Young's modulus.
- Poisson's ratio (v) is the ratio of the lateral strain to the axial strain.
- Shear modulus (G), also known as the modulus of rigidity. For isotropic materials, the shear modulus is a function of the modulus of elasticity and Poisson's ratio. For isotropic materials it is:

$$G = E/(2 \times (1 + v)) \qquad (3.2)$$

Further to the above material structural properties, the member cross-sectional properties must be established to analyse the structure or individual member. The most important cross-sectional, structural member properties are:

- cross-sectional area
- location of centroid i.e. centre of gravity
- moments of inertia
- section modulus

These plane area geometrical properties have a special significance in the numerous interactions which govern stress and deflection in structural members.

Due to the wide variety of aluminium cross-sectional shapes software is commonly used to calculate the required cross-sectional properties of members intended to be used in the structure. However, for standard symmetrical shapes these calculations can be handled manually and equations to determine section properties can be found in many compendia of structural and civil engineering data.

3.3.1   Cross-sectional area

The cross-sectional area (A), is perhaps the most basic property of the shape of a member. It is also the property that can be determined without difficulty for the vast majority of shapes that are used. Yet, the cross-sectional area is used to compute many other important cross-sectional properties, including: self-weight, the section moduli, the second moment of area and the radius of gyration.

For many basic shapes the cross-sectional area (A) is a simple function of the breadth multiplied by the depth of the shape. However, for non-symmetrical and arbitrarily shaped cross-sections, specialist software can be utilised to establish the cross-sectional area accurately. Many computer-aided design (CAD) software programs give areas for shapes when drawn to simple parameters of the specific software used. Figure 3.2 shows examples of cross-sectional areas for prismatic isotropic members.

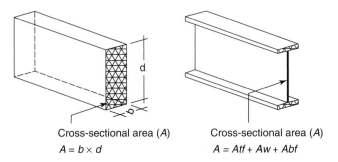

Cross-sectional area (A)   Cross-sectional area (A)

$A = b \times d$ $A = Atf + Aw + Abf$

Figure 3.2 | Cross–sectional areas for typical member shapes (shown hatched)

3.3.2   Location of centroid

The centroid or centre of gravity, of any object is the point within that object from which the force of gravity appears to act. An object will remain at rest if it is balanced on any point along a vertical line passing through its centre of gravity. In terms of moments, the centre of gravity of any object is the point around which the moments of the gravitational forces completely cancel one another out.

It is relatively simple to find the location of the centroid for singly or doubly symmetrical shapes. When a shape has one axis of symmetry, the centroid is located on that axis. When a shape has two axes of symmetry, the centroid of that shape lies at the intersection of the two axes. Figure 3.3 illustrates the location of the centroid on the axes of symmetry for singly and doubly symmetrical cross-sectional shapes.

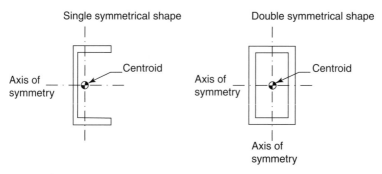

Figure 3.3 Centroid on axes of symmetry

3.3.3 Moment of inertia

The moment of inertia (*I*), or second moment of area is a measure of the manner in which this area is distributed about an axis of interest. It also indicates the resistance of a shape to bending and deflection and can be used to calculate the resistance to buckling and the state of stress.

The second moments of area of an area *A* about the *x*- and *y*-axes are given by (see Figure 3.4):

$$I_{XX} = \int_A y^2 dA \quad \text{and} \quad I_{YY} = \int_A x^2 dA \tag{3.3}$$

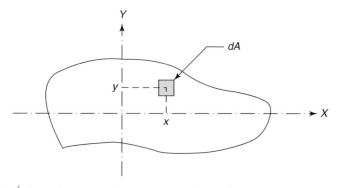

Figure 3.4 Second moment of area (moment of inertia)

Table 3.8 | Second moment of area for most common shapes

Shape	Second moment of area
	Square $\qquad$ **Rectangle** $I = \dfrac{a^4}{12}$ $\qquad\qquad$ $I = \dfrac{B \times H^3}{12}$
	Circle $\qquad$ **Circular tube** $I = \dfrac{\pi \times D^4}{64}$ $\qquad$ $I = \dfrac{\pi \times (D^4 - d^4)}{64}$
	Rectangular tube $I = \dfrac{(B \times H^3) - (b \times h^3)}{12}$
	H section $I = \dfrac{(B \times H^3) - (b \times h^3)}{12}$
	Channel/C section $I = \dfrac{(B \times H^3) - (b \times h^3)}{12}$
	Angle section/L section $I = \dfrac{(B \times e1^3) - (b \times c^3) + (a \times e2^3)}{3}$

Equations for the second moment of area of most common shapes can be found by applying direct derivations from the parallel axes theorem (see Table 3.8).

3.3.4 Section moduli

Further geometrical cross-sectional properties are the two section moduli, the elastic section modulus (W_{el}) and the plastic section modulus (W_{pl}). The section moduli are used to calculate the stress action in the extreme fibre of a shape when bending moments are introduced due to loading i.e. point load or distributed load on a beam.

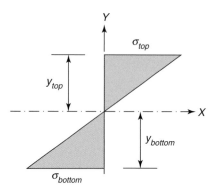

Figure 3.5 | Elastic bending stress diagram

Most manufacturers of aluminium prismatic bars will provide the relevant section moduli for the shapes of their bars.

Elastic section modulus

The elastic section modulus (W_{el}), is acquired by dividing the second moment of area (I) by the distance (y) measured from the principle axis of the section to the most outlying fibre of the section (see Figure 3.5):

$$W_{el} = I/y \tag{3.4}$$

For sections symmetrical about the major axis in bending, the elastic section modulus is the same for the top and bottom fibres since $y_{top} = y_{bottom}$. For asymmetrical sections the lesser elastic section moduli for the most extreme top (W_{top}) or bottom (W_{bottom}) is used in structural design.

Similar to the calculations of the second moment of area (I) the elastic section modulus can be generated directly from the geometric properties and dimensions of the cross-section. Equations for the elastic section modulus of most common shapes are listed in Table 3.9.

The effects of HAZs, holes to be drilled for connections and local buckling must be considered when computing the elastic section modulus. Therefore the accurate elastic section modulus should be generated using the section properties of the effective sections.

Plastic section modulus

The plastic section modulus (W_{pl}) is given by the first moment of area about the equal area axis. The plastic section modulus should only be used with compact or plastic sections.

Table 3.9 | Elastic section moduli for most common shapes

Shape	Elastic section modulus ($W_{el} = I/y$)
	Square $\qquad\qquad$ **Rectangle** $$W_{el} = \frac{a^3}{6} \qquad\qquad W_{el} = \frac{B \times H^2}{6}$$
	Circle $\qquad\qquad$ **Circular tube** $$W_{el} = \frac{\pi \times D^3}{32} \qquad W_{el} = \frac{\pi}{32} \times \frac{(D^4 - d^4)}{D}$$
	Rectangular tube $$W_{el} = \frac{1}{6 \times H} \times [(B \times H^3) - (b \times h^3)]$$
	H section $$W_{el} = \frac{1}{6 \times H} \times [(B \times H^3) - (b \times h^3)]$$
	Channel/C section $$W_{el} = \frac{1}{6 \times H} \times [(B \times H^3) - (b \times h^3)]$$
	Angle section/L section $$W_{el.1} = \frac{I}{e_1} \qquad W_{el.2} = \frac{I}{e_2}$$

Table 3.10 | Typical shape factors, ν

Section	Rectangle	I-section major axis	I-section minor axis	Solid tube	Circular hollow section	Channel section
Shape factor	1.5	1.15	1.67	1.7	1.27–1.4	1.15

A simple method to determine the plastic section modulus is to apply the shape factor to the elastic section modulus, $W_{pl} = W_{el} \times v$. The shape factor is dependent on the shape of the cross-section and typical values are given in engineering handbooks. A few are listed in Table 3.10.

In elastic structural design/analysis the maximum stress is found from the function:

$$\sigma_E = M/W_{el} \tag{3.5}$$

Similarly to the approach in elastic analysis, in plastic structural design, the maximum stress is determined from:

$$\sigma_P = M/W_{pl} \tag{3.6}$$

Due to the leverage about the equal area axis only one plastic section modulus is calculated, whilst the elastic section moduli for the top and bottom part of the section can be different (see Figure 3.6).

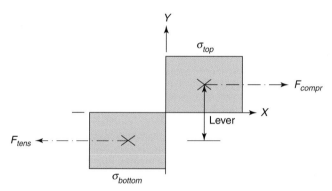

Figure 3.6 | Plastic bending stress diagram

3.3.5 Radius of gyration

The radius of gyration (i) is a further factor that is essential for structural design. The radius of gyration is used to describe the distribution of cross-sectional area around its centroidal axis. The value for the radius of gyration is calculated from:

$$i = \sqrt{(I / A)} \tag{3.7}$$

The radius of gyration is practical for buckling analysis and indicates the resistance to buckling of a cross-section. A distributed area further away from the centroidal axis will result in a greater value for the radius of gyration and equally better resistance to buckling of the section.

Example 3.3: Comparison of two cross-sections with equal cross-section area:

Section 1:   SHS 100 × 100 × 10
Section 2:   Bar 60 × 60

Equations for solution:

cross-sectional area, A = S1: $a^2 - (a - 2t)^2$ | S2: a^2
moment of inertia, I = S1: $[a^4 - (a - 2t)^4] \div 12$ | S2: $a^4 / 12$
radius of gyration, $i = \sqrt{(I/A)}$

Solution:

Section	1	2	Units
area, A	36	36	cm²
$I_y = I_z =$	492	108	cm⁴
$I_y = i_z =$	36.9	17.3	mm

Thus section 1 with the same area, but distributed further away from the centroidal axis has a greater resistance to buckling than the narrower solid bar. This can be seen instantly from the value for the radius of gyration.

3.4 Structure and element analysis

The theories for structural analysis, beam bending and structural mechanics will not be considered in this volume. Thus, only the basic requirements are considered in this chapter. The reader should be familiar with the principles of structural engineering, structural analysis and structural mechanics.

Before member design can commence, the structure or single member must be analysed to determine the member actions and displacements. This process requires loadings and member properties to be applied to each member, often called creating the structural model. Structural models can be made up of one member, or several members, as seen in frames or trusses. With the use of modern analysis software three-dimensional models are frequently applied to find the interaction between the members of the models in all dimensions.

When creating structural models, the usual procedure involves producing the structural elements or members with their geometrical properties and applying the relevant mechanical properties and loadings. It is essential to understand this process and its restrictions, since geometry and member properties of the model are generally idealised. Hence, these models are, to a point, imperfect but can still result in an appropriate estimate for the behaviour of the structure. The engineer's experience, in combination with a suitable structural model, will provide suitable, safe and economical solutions.

Structural models can be made of 'stick' or 'line' structures, 'plated' or 'shell' structures and 'mass' structures. Line structures are made of line elements and are most commonly used in structural analysis. Line elements represent structural members such as: columns, beams, rafters, struts, braces and ties. Geometric and mechanical properties are applied to these line elements. Examples of line structures are illustrated in Figures 3.7–3.9.

The structural analysis is then carried out by calculation methods for each member, predicting the behaviour of the single member and/or the entire structure. This detailed study of the structural model will give results such as the deformation and member actions, required for structural design i.e. determination of sufficient sizes for the members of the structural model.

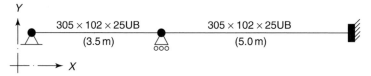

Figure 3.7 | Simple two-dimensional structural model

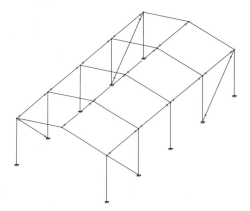

Figure 3.8 | Simple three-dimensional structural model/portal frame

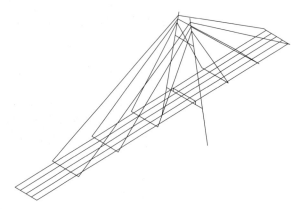

Figure 3.9 | Three-dimensional structural model of a footbridge structure

The concept of static equilibrium is fundamental to the analysis of a structure. The static equilibrium is present where a structure i.e. structural system subject to static loads is initially at rest and continues to be at rest when external loads are applied. This fact is based on Newton's law, which states that for every action there is an equal and opposite reaction. From observation it can be said that if a structural system is in equilibrium, then any component of the structural system is also in equilibrium.

A structure is in equilibrium if all the following conditions are satisfied:

$$\Sigma F_x = 0, \ \Sigma F_y = 0, \ \Sigma F_z = 0, \ \Sigma M = 0$$

The results of structural analysis can be listed numerically, in tabular form or graphically with diagrams for bending moments, shear forces and displacements.

Whilst only a little guidance for the design engineer is given in BS 8118: Part 1, the new Eurocode 9 (BS EN 1999–1–1:2007, Chapter 5) addresses the structural analysis and provides advice for the structural designer.

3.5 *Deformation*

Elasticity, the ability of a material to return to its original shape after the removal of all loading, is an important property in structural engineering. By this definition aluminium is an elastic material. The degree of elasticity of a material is given by its modulus of elasticity, represented by the symbol E and given by E = stress/strain. A higher modulus of elasticity indicates a more brittle material, whilst a smaller modulus of elasticity presents a more ductile material. A few moduli of elasticity are given in Table 3.11.

With the decreasing modulus of elasticity it becomes more likely that deformation will be the dominant design criteria rather than resistance to bending moment or shear forces. However, the displacements of the structural members are also directly dependent upon their support conditions, span and pattern of loading. Equations to compute the displacement for specific structural conditions are listed in various compendia of engineering formulae. Whilst the deformations can be adequately calculated by hand, it is appropriate to use

Table 3.11 | Modulus of elasticity for elastic materials

Material	Mild steel	Cast iron	Aluminium	Timber
E-modulus	205 kN/mm²	130 kN/mm²	69 kN/mm²	10 kN/mm²

Table 3.12 | Typical limiting deflections

Structural element	Defection limit
Cantilever beam	$0.005 \times$ span
Beams	$0.003 \times$ span
Columns (at top)	$0.003 \times$ height
Purlins	$0.005 \times$ span
Curtain wall elements	$0.004 \times$ span

either computer-based spreadsheets or software to assess the expected displacement under applied loading and conditions.

As excessive deformation would influence structural integrity, appearance and the well-being of the users of a structure, displacement limits are specified in the serviceability clauses of BS 8118 and also Eurocode 9. The recommended deflection limits shown in Table 3.12 are only generic. Further advice must be sought in the relevant clauses of BS 8118 and Eurocode 9.

Excessive deformation of a structure should be avoided as this can greatly influence the analysis results and stability of the structure. If the effects of the deformed structure increase the member actions or modify the structural behaviour, second-order analysis should be carried out. The second-order analysis (non-linear) analysis takes into account the second-order effects due to displacement or rotation. Eurocode 9 states a factor $\alpha_{cr} \geq 10$ for neglecting the influence of second-order effects: where α_{cr} is the result of the fraction of critical elastic buckling load for global instability (F_{cr}) divided by the design loading for the structure (F_{Ed}):

$$\alpha_{cr} = F_{cr}/F_{Ed} \geq 10 \qquad (3.8)$$

3.6 Member actions

Aside from the computation of the deformation of a structure, further results and information from the analysis of a structural model include the member actions. Member actions can be shown for all load cases and combinations using characteristic but also factored structure and member load. Again, the determination of the member actions can be prepared by hand calculations or computerised numerical analysis.

Results for member actions such as bending moments, shear and axial forces and displacement are often displayed in diagrams for each member separately, in preparation for member design.

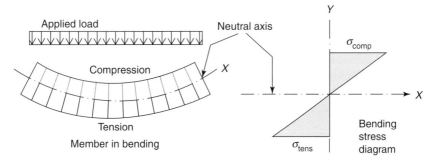

Figure 3.10 | Principle of member bending

3.6.1 Bending moment

A bending moment exists in a structural member, when an external action is applied to this member causing the member to bend. Tensile and compressive stresses are created by the bending action within the member (see Figure 3.10). The values of the bending moment are calculated by multiplying the applied external force by the distance at which these forces are applied.

For member design the extreme or design values are annotated within the bending moment diagram, however, the maximum values can also be isolated by spreadsheet or software filters. The values for bending moment member actions are given in units of Newton-metres (kNm or Nm). A typical bending moment diagram is shown in Figure 3.11 which illustrates a bending moment diagram for a uniformly loaded (5 kN/m) and simply supported beam giving a bending moment of 5.625 kNm at mid-span and no bending moment at the supports.

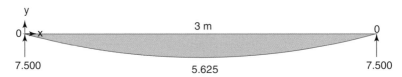

Figure 3.11 | Bending moment diagram (BMD)

3.6.2 Shear force

A structural member subject to bending actions also must resist the acting shear forces. In loaded elements where shear forces coexist with bending moments, they cannot be neglected in structural design (see Figure 3.12). Pure shear, as found in bolts or rivets, is less often found in beams.

However, shear also exists on other longitudinal surfaces and where on a given surface the shear stress can be assumed to being constant.

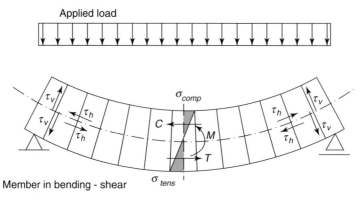

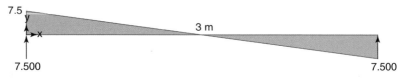

Figure 3.12 | Principle of member bending: vertical and horizontal shear

Figure 3.13 | Shear force diagram (SFD)

The longitudinal shear force (T), due to the shear force (V), passing through the shear centre of a typical structural member and is given by:

$$T = (V \times Q)/I \qquad (3.9)$$

where Q = first moment; V = shear force; and I = second moment of inertia. The critical position for shear force on a normally loaded flexural member is at the support where the maximum reaction occurs. Figure 3.13 illustrates a shear force diagram for a uniformly loaded (5 kN/m) and simply supported beam giving a maximum shear force of 7.5 kN at the beam supports.

3.6.3 Axial force

Axial forces, such as compression or tension, can exist on their own or in combination with bending actions. The stress distribution due to axial forces is assumed to be constant on the affected surface. When axial forces are found acting together with bending actions (shear and moment), it is necessary that the separate effects of these forces be combined to give the combined design stress.

3.6.4 Support reaction

Often neglected in structural design, the support reactions of members occupy an important role in the structural design of members. They are especially important when using thin-walled members which can easily be crushed or buckle due to the load concentration at the supports.

Computation of the support reactions is a straightforward task for simple structural systems and models. Structural engineering handbooks and formula reference books provide a good source of equations for calculating the support reactions manually. However, for more complex or three-dimensional models the aid of sophisticated computerised tools (spreadsheets or software) will result in accurate support reactions.

3.7 Finite element analysis

Finite element analysis (FEA) is commonly used to analyse elements/members in structural engineering. Due to advances in computing and increasingly powerful but also affordable computers, FEA has developed into the preferred tool for detailed analysis. It produces diagrams that are different from those described earlier in this chapter for other methods. Figure 3.14 shows a typical FEA diagram for stress within an aluminium plate. Discussing the principles of FEA is beyond the scope of this book. However, it is the intention to illustrate different analysis result diagrams based on the FEA method and software used.

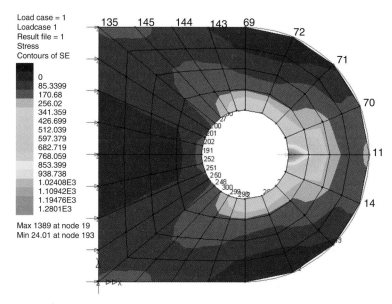

Figure 3.14 | Example of analysis results using FEA

4

SERVICEABILITY LIMIT STATES

Introduction

Serviceability refers to the states under which a structure is considered suitable for use. Serviceability requirements might fail a structure, and thus classify the structure or building as unfit for its intended use despite all the members of the structure being structurally safe and complying with the relevant structural design codes.

Thus, the serviceability of a structure or building addresses its function, use and appearance in association with the purpose and user requirements. Other serviceability criteria include its coexistence with other installations and the fitment of other materials to the structure or building in order to prevent damage or loss of the required function.

Low modulus of elasticity but good strength capacities will often prove to be problematic when checking the global and/or local displacements of aluminium and aluminium alloy members and structures against the requirements of the serviceability limits given in the relevant design codes.

However, smart design which exploits the strengths of the material in tension and compression, while limiting exposure to bending actions, will give high-strength, good quality structures made from aluminium materials, which satisfy the SLS requirements.

SLS structural design checks include: computation of displacements, overall stability, vibration and fire resistance. The serviceability limit specifies requirements for the comfortable use of a structure or building for deflection and vibration, and also its overall stability and behaviour in fire situations. A structure will fail the serviceability criteria when it exceeds the given or agreed limits for one of the following states:

- overall deflection of entire structure (global displacement)
- member/element displacements (local displacement)
- vibration
- fatigue

Table 4.1 | Load factors for SLS

	Eurocode	BS 8118
Permanent action(s)	1.00 G_k (g_k)	1.00
Variable action (leading)	1.00 Q_k (q_k)	1.00
Variable action(s) (others)	$\psi_{0/1/2} \times Q_k$ (q_k)	1.00

The limits for serviceability with respect to structural aluminium design are laid out in BS 8118, BS EN 1990 and BS EN 1999. The limits for serviceability are usually checked against the characteristic loads i.e. unfactored loads or reduced loads in special circumstances (see Table 4.1).

4.1 Displacements

Displacements occur in structures and structural elements when loaded vertically or laterally. However, temperature changes within the member or element resulting in contraction or expansion, may also cause displacement of the members.

Deflection beyond specified limits can seriously affect the use and function of a structure or building. Therefore, excessive deformation should be avoided and displacement must be checked against limits given in the relevant standards and codes.

With a lesser modulus of elasticity ($E = 69$ kN/mm²) than steel, the deflection of aluminium elements can be the dominant design criteria with bending, shear and axial stresses still within allowable limits. Special care must be taken adjacent to the areas which have been identified as being affected by HAZs. Due to the softening of the material, greater displacements are possible in these areas. These displacements at HAZs can develop irreversible plastic deformations and need to be checked separately from the elastic and reversible displacements. In both cases, the limits specified in the British Standards and Eurocodes should be rigorously applied. Displacement limits to Eurocode 9 are given in the National Annexes to BS EN 1990. Further guidance and deflection limits can be found in Annex A.1.4 of BS EN 1990.

Suitable deflection limits can also be agreed between the designer and user of the structure. This departure from the design codes can give a more efficient solution, with allowable deflection limits based on the functionality of the structure. This procedure is often applied to structures with a special function or moveable parts where elements interact. A typical example is a substation structure carrying an electrical 'disconnector' with movable blades that slot into fixed parts of the 'disconnector' equipment. Surpassing agreed displacement limits

Table 4.2 | Formula for computation of displacement

Member/loading condition	Maximum displacement(*d*)
Cantilevered beam/PL	$d = \dfrac{G \times L^3}{3 \times E \times I}$ at beam end
Cantilevered beam/UDL	$d = \dfrac{g \times L^4}{8 \times E \times I}$ at beam end
Simply supported beam/PL	$d = \dfrac{G \times a^2 \times b^2}{3 \times E \times I \times L}$ at point of load
Simply supported beam/UDL	$d = \dfrac{5 \times g \times L^4}{384 \times E \times I}$ at mid-span
Beam fixed ends/UDL	$d = \dfrac{1 \times g \times L^4}{384 \times E \times I}$ at mid-span
Beam fixed end - pinned end /UDL	$d = \dfrac{2 \times g \times L^4}{369 \times E \times I}$ at $x = 0.579\,L$

could result in the blades missing the pockets they should slot into, leaving the equipment failing to achieve the required function.

Some of the typical equations which are used to calculate possible displacements for the most common beam types and loading conditions are given in Table 4.2. It should be noted that deflection calculations should be performed using section properties which allow for local buckling of slender sections. Typical deflection limits are given in Table 4.3.

The consequences of an excessive displacement of a structure or its single elements are:

- discomfort of occupants and users
- reduction or total loss of function
- damage to secondary fitments and finishes
- damage to structure or single member (plastic deformation)
- second-order effects occur leading to structural failure

4.2 Vibrations

Vibrations, also part of the dynamic response of structures or structural members, are generated by dynamic loads acting in the structure.

Table 4.3 | Typical deflection limits

	Vertical	**Horizontal**
Total deflection	1/250 × span	
Live load deflection	1/360 × span	
Deflection of brittle elements	1/500 × span	
Cantilevers	1/180 × span	
Purlins (dead load only)	1/200 × span	
Sway of single-storey columns		1/300 × height
Sway of multi-storey columns at each storey		1/300 × height
Sway of columns with movement sensitive cladding		1/500 × height
Curtain wall mullions and transoms (single glazed)		1/175 × span
Curtain wall mullions and transoms (double glazed)		1/250 × span

However, whilst many dynamic loads can easily be identified, some are more concealed within the 'imposed' or 'live' load category, such as walking occupants or other movable *live* loads. Wind loads are usually found within the static load cases, but when they act on tall or slender structures they can introduce damaging aerodynamic effects and need to be taken into consideration.

Typical sources of dynamic loads (vibrations) include:

- walking and synchronised movement of people (floors and bridges)
- traffic vibrations from nearby roads and railways
- aerodynamic effects such as wind load, wind buffeting etc.
- fitted equipment and machinery
- construction activities, e.g. piling

Only a small amount of guidance on dealing with the dynamic response and vibration of structures and structural elements is given in BS 8118 and Eurocode 9. Nevertheless, if susceptible to dynamic response, structures should be checked accordingly and in more detail to prevent unfavourable responses. In the absence of official regulation, the sources of vibration and serviceability limits should be specified and agreed with the client and/or user of the structure.

The design engineer should check whether or not the structure or structural member is subjected to induced vibration at its natural frequency. If the structure is exposed to vibrations at its natural frequency, the displacements (i.e. vibrations) will develop to their maximum rate resulting in discomfort to the occupants, loss of function or damage to the structure (see Table 4.4).

Table 4.4 | Formulae for natural frequencies

Member/loading condition	Natural frequency(f)
L = beam length M Beam of negligible mass Cantilevered beam/PL	$f = \dfrac{1}{2 \times \pi} \sqrt{\dfrac{3 \times E \times I}{M \times L^3}}$
L = beam length m = beam mass/unit length Cantilevered beam/UDL	$f = \dfrac{k}{2 \times \pi \times L^2} \sqrt{\dfrac{E \times I}{m}}$ Mode: 1, 2, 3, 4 — k: 3.52, 22.0, 61.7, 121.0
M -$L/2$- -$L/2$- -L- Beam of negligible mass Simply supported beam/PL	$f = \dfrac{1}{2 \times \pi} \sqrt{\dfrac{48 \times E \times I}{M \times L^3}}$
M -a- -b- -L- Beam of negligible mass Simply supported beam/PL	$f = \dfrac{1}{2 \times \pi} \sqrt{\dfrac{3 \times E \times I}{M \times a^2 \times b^2}}$
-L- Simply supported beam/UDL	$f = \dfrac{k}{2 \times \pi \times L^2} \sqrt{\dfrac{E \times I}{m}}$ Mode: 1, 2, 3, 4 — k: 9.87, 39.5, 88.8, 158.0
-L- Beam fixed ends/UDL	$f = \dfrac{k}{2 \times \pi \times L^2} \sqrt{\dfrac{E \times I}{m}}$ Mode: 1, 2, 3, 4 — k: 22.4, 61.7, 121.0, 200.0
-L- Beam fixed ends - pinned end/UDL	$f = \dfrac{k}{2 \times \pi \times L^2} \sqrt{\dfrac{E \times I}{m}}$ Mode: 1, 2, 3, 4 — k: 15.4, 50.0, 104.0, 178.0

The mode/k sub-tables for the rows above are:

Cantilevered beam/UDL

Mode	1	2	3	4
k	3.52	22.0	61.7	121.0

Simply supported beam/UDL

Mode	1	2	3	4
k	9.87	39.5	88.8	158.0

Beam fixed ends/UDL

Mode	1	2	3	4
k	22.4	61.7	121.0	200.0

Beam fixed ends - pinned end/UDL

Mode	1	2	3	4
k	15.4	50.0	104.0	178.0

The consequences of excessive vibrations introduced to a structure or its single elements include:

- discomfort to occupants and users
- reduction or total loss of function
- damage to secondary fitments and finishes
- fatigue

4.3   Damage

Like SLS failures due to deflection and vibration, damage to a structure or structural members can have severe effects on the structural and functional performance of a structure or structural member.

The new Eurocodes include damage which causes adverse effects on the appearance, durability and function of the structure. Thus, the designer of the structure should consider methods and procedures to reduce the risk of damage during fabrication, transport, erection and commissioning of the structure.

No detailed guidance is given and no specific requirements are listed in BS 8118 and Eurocode 9, as to the precautions which need to be in place to prevent damage, or reduce the risk of damage to a structure. Structural members are most exposed to accidental damage or overstressing during fabrication and erection, during transport, and when being loaded and offloaded.

Design checks should be carried out on the lifting of: vulnerable members which are weak about the minor axis, slender and heavy elements, and elements where the lifting points do not match with the intended support locations. The design engineer must specify safe lifting points for those members, elements and structures. Erection sequencing and method statements can further reduce the risk of damage during erection and provide the erecting team with essential guidance for their work.

5

ULTIMATE LIMIT STATES

Introduction

Ultimate limit states (ULSs) design refers to the states under which a structure is considered safe for its intended use. Hence, it considers the structural design that will give structures that will not present a danger to the users, the structure or parts of the structure. Like the SLS design, checks carried out against the ULSs design criteria can fail a structure or parts of it despite it having satisfactorily met the SLS requirements.

The ULSs design requires the structural designer to check the design resistance of the structure against applied design actions/design loadings. But structural equilibrium or excessive deflections beyond the elastic limits must also be examined as part of ULS design checks. Failure, i.e. reaching the ULS, would mean the probability of collapse or non-reversible damage to the structure or part of it.

Other than with serviceability design checks, loadings and material values are factored by partial safety factors and load factors, respectively. By using this factoring process, loadings can be multiplied and material values can be divided by the appropriate factors, resulting in the design values as specified by the relevant design codes. Both design codes BS 8118 and BS EN 1999 (Eurocode 9) are limit states design codes.

5.1 Principle of ultimate limit state design

To satisfy the requirements of ULS design the structure or individual parts of the structure must have design resistance to all applied design loadings/design action analysed. This basic but all important condition is summarised in Equation (5.1):

$$\text{Design action} \leq \text{design resistance}$$

$$F_d / R_d \leq 1.0 \tag{5.1}$$

where F_d = design action; and R_d = design resistance. The above condition should be checked against all possible failure modes! It must be fulfilled to pass the strength design checks.

Further to the design for static strength, it may be necessary to establish the actual factor of safety against total failure or collapse. This can be calculated as:

$$FoS_{\text{collapse}} = R_k/F_k \qquad (5.2)$$

where

FoS = factor of safety
F_k = characteristic action
R_k = characteristic resistance

The characteristic resistance is calculated by:

$$R_k = R_d \times \gamma_{Mi} \qquad (5.3)$$

where: γ_{Mi} = partial safety factor(s).

5.2   Design values

Since ULSs are directly linked to possible structure failures i.e. the collapse of the entire structure or parts of the structure, ULSs design should certify a low likelihood of occurrence. In order to achieve increased reliability and lesser possibility of failure, material and load factors are used to account for unknown aspects within the structural design. By using the above-mentioned material and load factors it should be guaranteed that an adequate factor of safety is given against reaching or surpassing the ULSs.

In view of the impending implementation of Eurocode 9, the material and load factors given below are on the basis of guidance and recommendations given within the Eurocode 9 (BS EN 1999).

5.2.1   Material factors/partial safety factors

The current theory of structural safety and reliability is based on a semi-probabilistic verification method. The uncertainties in the accuracy of factors affecting structural design are addressed by the application of reduction factors, the partial factors of safety. These factors should not be mistaken for load factors, which will be discussed in Section 5.2.2. For some time code developers and code authorities have aimed to harmonise the structural design codes at international level in an attempt to achieve a homogeneous level of structural safety and reliability in structural design. The Eurocodes family is a result of this harmonisation process.

Partial safety factors were already incorporated in the design guidance given by BS 8118 and are an important part of the structural

Table 5.1 | Typical partial safety factors and EC 9 notation for ULSs

Design category	PSF γ_{Mi}	EC 9 notation
Resistance of members	1.10	γ_{M1}
Resistance in tension (to fracture)	1.25	γ_{M2}
Resistance of bolted joints	1.25	γ_{M2}
Resistance of riveted joints	1.25	γ_{M2}
Resistance of pinned joints	1.25	γ_{M2}
Resistance of welded joints	1.25	γ_{M2}
Resistance of bonded joints	3.00	γ_{Ma}
Slip resistance of joints	1.25	γ_{M3}
Bearing resistance of plates	1.25	γ_{M2}
Preload of high strength bolts	1.10	γ_{M7}

Bonded = adhesively bonded (glued)

design method described in BS EN 1999–1-1. Typical partial safety factors for ULSs design are illustrated in Table 5.1.

When using partial safety factors in structural design to Eurocode 9 (BS EN 1999), the values of γ_{Mi} may be defined differently in the National Annexes or clauses of the Eurocode 9. Further guidance is given in BS EN 1999 and also in BS 8118.

The principle for utilisation of the partial safety factors is expressed in Equation (5.4) to determine the design resistance R_d:

$$R_d = R_k / \gamma_{Mi} \tag{5.4}$$

where

γ_{Mi} = partial safety factor(s)
R_d = design resistance
R_k = characteristic resistance

5.2.2 Load factors/load combinations

A first insight into loadings and load factoring was presented in Section 3.2 of this book where the basic principles of loading (actions) and application of the relevant load factors were demonstrated. In contrast to the current code BS 8118, where load application, load combination and associated load factors are specified, the Eurocode 9 is for design guidance only and actions and applicable factors are dealt with by Eurocode 0 (BS EN 1990). Thus the actions and factors for deriving design values for actions are harmonised over the entire Eurocode family and are independent of the structural material and related design codes.

Whilst in the versions of British Standards the combinations of loads applied to a structure have conveniently been called 'load combinations', the Eurocode 0 (BS EN 1990) describes combined actions as 'design situations'. Despite a new name, these design situations are nothing but specific load combinations and are classified as:

- persistent design situations (permanent i.e. during normal use)
- transient design situations (temporarily e.g. during erection)
- accidental design situations (fire, explosion and impact)
- seismic design situations (all seismic events)

For the majority of structural design at ULSs two design situations will predominate: the persistent design situation and the transient design situation. The design situation selected for structural design must be sufficiently rigorous and diverse to reflect the conditions which can be expected during the design life of a structure. For the purpose of this book, only the persistent design situations are depicted, in that they will be the most frequently employed design situation for strength design checks.

BS EN 1990 (Eurocode 0) specifies the calculation method for the design actions and design situations for ULSs design in clauses 6.3 and 6.4. Design values for actions are obtained from:

$$F_d = \gamma_f \times \psi_i \times F_k \tag{5.5}$$

where

F_d = design value of action
F_k = characteristic action
ψ_i = either 1.0, ψ_0, ψ_1, or ψ_2
γ_f = partial factor for action(s)

Design values for the design situation are given by:

Sum of design permanent actions + leading design variable action + plus sum of other design variable action(s)

This is best illustrated by Equation (5.6):

$$\sum_{i \geq 1} \gamma_{G.i} \times G_{k.i} + \gamma_{Q.1} \times Q_{k.1} + \sum_{i > 1} \gamma_{Q.1} \times \psi_{0.i} \times Q_{k.i} \tag{5.6}$$

where

γ = partial factor for action(s)
ψ = combination factor(s)
G_k = permanent action(s)
Q_k = variable action(s)

5.3 Failure states

For ULSs design, the limit states listed below must be verified if they are relevant to the structural design:

- loss of static equilibrium
- loss of structural strength
- failure due excessive deformation of ground/foundations
- fatigue failure

It is the structural designer's task and responsibility to ensure that a structure or components of a structure are performing in a satisfactory manner, i.e. are safe to use, and that the above-mentioned states and associated actions on the structure or individual structural member are analysed and addressed. A satisfactory design will demonstrate a structural capacity greater than the applied design actions.

5.3.1 Loss of static equilibrium

The principle of static equilibrium is based on Newton's law, where every action must have an equal reaction. Therefore, a structural system is in equilibrium when every element of a structure, including the joints, is in static equilibrium. Failure, i.e. loss of static equilibrium, will have serious consequences, if not total failure and collapse of the structure.

Static equilibrium can be summarised by the expression 'sum of all forces' equals 0, but also 'sum of all moments' equals 0.

$$F = 0, \quad \Sigma M = 0 \tag{5.7}$$

The see-saw is a simple example of a structural system. It will only come to rest if the above conditions are met, hence static equilibrium is obtained. An illustration of the principle of static equilibrium applied to the see-saw structure is shown in Figure 5.1.

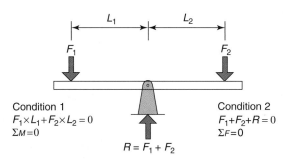

Figure 5.1 | Static equilibrium

If one of the applied forces (F_1 or F_2) is increased against the other, the see-saw beam would start to rotate until stopped by the ground. The ground will then provide additional support, restoring static equilibrium to the structural system.

5.3.2 Loss of structural strength

Loss of structural strength is considered an internal failure, where a single or multiple members of a structure can fail the strength requirements due to insufficient resistance to applied actions. This includes excessive deformations but also connection failures.

Loss of structural strength can also occur when a structure is exposed to severe conditions or a harsh environment. Examples include structural behaviour in fire or exposure to corrosion attack where a structure, members or connections can lose structural strength over time after passing initial structural calculations.

It is vital that all members and connections forming a structure are checked against the requirement:

Design Action ≤ Design Resistance

5.3.3 Failure due excessive deformation of ground

The majority of structures will require support from the ground. Exceptions are planes when in flight and ships floating in water. In most cases, failure or excessive deformation will result in loss of static equilibrium and the collapse of a structure. Unless the static equilibrium is restored after ground failure or ground deformation has occurred.

The most infamous building is the Leaning Tower of Pisa in Italy (see Figure 5.2). Part of the footings of the tower was set in soft and

Figure 5.2 | Leaning Tower of Pisa

unstable ground which resulted in the tower already leaning during construction.

Guidance and the design requirements for geotechnical design are given by Eurocode 7 (BS EN 1997) and its parts 1 and 2.

5.3.4 Fatigue failure

Fatigue failure occurs when structural members are subjected to cyclic and changing loading and deformation. Fracture often takes place suddenly and at stresses below the ultimate design resistance. Who has not bent a wire or metal sheet continuously until it broke?

Good design and fabrication techniques are necessary to prevent damage due to fatigue. Detailed guidance is given in BS 8118: Part 1: Chapter 7 Fatigue, and Eurocode BS EN 1999–1-3 Structures susceptible to fatigue.

Figure 5.3 shows fatigue curves which illustrate the typical fatigue behaviour of aluminium compared to steel. Whilst steel has the unique property of a fatigue limit at 10^7 cycles where no fatigue failure would occur, aluminium possesses no fatigue limit and will continue to be vulnerable to fatigue failure.

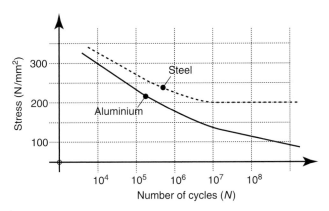

Figure 5.3 | Fatigue curves for aluminium and steel

6

FATIGUE

Introduction

Aluminium members and their connections which are subjected to cyclic loading or deformations can fail suddenly due to brittle fracture at a grade below their usual material strength. This failure is known as fatigue.

Because of this possibility of sudden and unexpected failure at low stress levels, fatigue considerations are an important part of the structural design process. Fatigue design is a very complex subject requiring a high level of skill from both the designer and the fabricator. The importance of fatigue design is reflected in the expanded design guidance given by Eurocode 9, which has a separate part addressing structures susceptible to fatigue (BS EN 1999, Part 1–3).

Despite the expanded guidance that is available, the identification of susceptibility to fatigue is vital in order to instigate design and other measures to reduce the risk of fatigue failure, or at least increase the life expectancy of a structure.

Often, simple fabrication actions such as grinding, weld toe re-melting, overloading and peening can greatly reduce vulnerability to fatigue. However, careful execution is required as exceeding the specified preventative measures could result in further weakening of the structural members and/or connections.

The process of fatigue can be identified by its three phases:

1) crack initiation
2) crack propagation
3) fracture

6.1 Fatigue susceptibility

It is important to identify the fatigue susceptibility of a member to allow for design against fatigue. A number of factors can either lead to or benefit fatigue crack growth. Once this has been recognised and classified, preventative measures can be utilised to limit the risk of fatigue failure.

Conditions for fatigue susceptibility
Conditions for fatigue susceptibility are listed and explained in the design codes and are summarised below:

- high ratio of dynamic to static load
- frequent/repeated application of load
- use of welding
- complexity of joint details
- environmental conditions

Types of loading
Types of loading that could lead to fatigue cracks and fatigue failure are:

- moving loads (on bridges, cranes and machinery)
- vibrations (machinery and vehicles)
- environmental loadings (wind and wave loads = changing loads)
- repeated pressurisation (machinery, vehicles, pipes and vessels)
- repeated temperature changes

Possible locations of fatigue initiation
The most common locations where fatigue cracking can occur are listed below and typical situations are shown in Figure 6.1:

- toes and roots of fusion welds
- machined corners and drilled holes
- surfaces under high contact pressure (fretting)
- roots of fastener threads
- sudden changes in section size/shape

6.2   Design against fatigue

Design against fatigue requires experience and a high level of skill in structural and material engineering in order to determine the fatigue

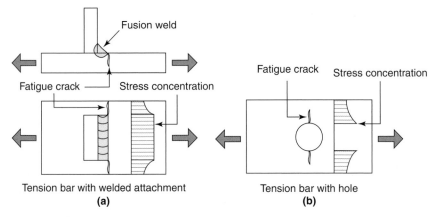

Figure 6.1 | Typical sites for initiation of fatigue cracks; (a) at welds; (b) at drilled holes

limits of structural elements. If inspection shows that a structure or member is vulnerable to fatigue, there are various approaches to predict the lifespan of a part and increase the fatigue resistance or replace the affected part at a given lifespan. The main methods to appraise a structure or single part of a structure and ensure structural integrity are:

- Structural design to reduce stresses below the fatigue limits.
- Design for a defined lifespan i.e. safe life with the specified part or structure to be replaced after the specified period has expired.
- Thorough inspection, testing and maintenance of a structure or part. A part of the structure is replaced after damage or cracks have occurred which exceed a predefined level/length.

Detailed guidance is given in BS 8118: Part 1: Chapter 7 Fatigue, and Eurocode BS EN 1999–1-3 *Structures susceptible to fatigue*. In-depth discussion of the methods described in these design standards is beyond the scope of this book. However, it is the author's intention to reveal the significance of fatigue effects in aluminium alloy structures and aluminium alloy parts. As, when overlooked, structures at risk to fatigue cracking can collapse suddenly with costly consequences to users and owners likewise.

Nevertheless, it is of equal importance to know about some basic measures to reduce the risk of fatigue cracking. The main improvements are given in Section 6.3 *Measures to limit formation of fatigue*.

Although the stated improvements are methods in manufacture to improve the fatigue performance of a part, joint or member, the design engineer should be aware of these methods in order to specify as appropriate and required. These manufacturing methods have been discussed in detail by Gurney (1979).

The fatigue strength of a member or welded joint is greatly influenced by the quality of the workmanship and welds. It is therefore imperative to know the limitations of the fabricators prior choosing the fatigue strength for structural design.

6.3   Measures to limit formation of fatigue

6.3.1   Grinding

Grinding of weld toes and weld ends will reduce the risk of fatigue cracks forming. This can be done by grinding grooves in the direction of the member (burr grinding) or grinding grooves transverse to the member (disc grinding). The lower stress concentration factor and the removal of crack-like defects at the weld toe generally give

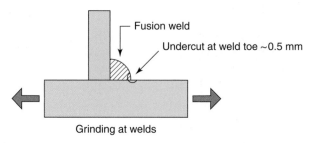

Figure 6.2 | Undercutting at weld toe

large increases in fatigue life. It is vital that the toe material of the weld is removed (see Figure 6.2). To ensure removal of the slag intrusions, a minimum depth of 0.5–1.0 mm is recommended. But it is vital not to remove too much material, which would further weaken the main part.

6.3.2 Tungsten inert gas dressing (weld toe re-melting)

Tungsten inert gas (TIG) welding can be used to melt the weld toes and weld ends. The surface of the weld should be cleaned by brushing before TIG dressing (see Figure 6.3). The TIG dressing method produces an improved weld profile, thus reducing the stress concentration at the weld toe due the smoother weld toe transition. But slag inclusions and undercuts are also removed.

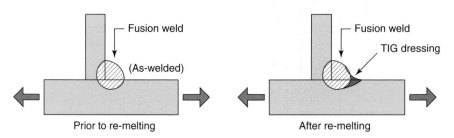

Figure 6.3 | TIG dressing (re-melting of weld toe)

6.3.3 Overloading

Overloading is used to introduce compressive stresses which are beneficial at possible fatigue locations. This is done by giving the member with possible fatigue sites a controlled static overload to a stress level beyond the elastic limit. Compressive residual stresses will not initiate fatigue cracking and will increase the fatigue strength of a member.

6.3.4 Peening

Peening releases the tensile stresses in the weld by cold-worked stretching. This method introduces local compressive residual stresses, thus improving the fatigue resistance. Various methods such as: hammer-peening, shot-peening, needle-peening, rotary-flap-peening and brush-shot-peening, are widely used to prevent fatigue crack initiation at weld toes. Hammer-peening and shot-peening have been proven to be highly effective methods for improving the fatigue strength of welded connections.

References

Gurney T R 1979. *Fatigue of Welded Structures*, 2nd edn., Cambridge University Press, Cambridge, UK.

7

STRUCTURAL MEMBER DESIGN

Introduction

Structural member design can be most easily described by comparing the design resistances of a member or element against the design actions that are being analysed. It is a repetitive member-by-member process that extends to the entire structure that is being addressed, whether this is a single beam or a comprehensive two- or three-dimensional framework. It aims to demonstrate the structural adequacy of each member of the examined structure.

It should be noted that the design of associated connections is often done separately, possibly by a different structural designer. It is important to check that weakening by joint fabrications, i.e. drilling holes, welding or added eccentricity will not reduce the member capacity below the required member resistance. In some cases it may be necessary to increase the size of a member section to suit the connection requirements.

This chapter describes the design methods and procedures on the basis of the guidance given in the European Standard for structural design of aluminium structures, Eurocode 9 (BS EN 1999), which is due to be fully implemented in 2010. For ease of understanding, and in keeping with the introductory purpose of this book, the methods portrayed are kept as simple as possible. Some basic examples have been used to illustrate the design checks and related procedures that are required.

7.1 Design standards: structural design codes

7.1.1 British Standard 8118 (BS 8118)

The British design code for aluminium structures in the UK is British Standard 8118. It has two parts:

- BS 8118: Structural use of aluminium, Part 1: Code of practice for design
- BS 8118: Structural use of aluminium, Part 2: Specification for materials, workmanship and protection

7.1.2 Eurocode 9 (BS EN 1999)

The new Eurocode 9 deals with the structural design of aluminium structures. All the parts of Eurocode 9 have been published and should replace the current British Standard 8118 by 2010. Eurocode 9 consists of five parts:

- BS EN 1999: Part 1–1: General, common rules
- BS EN 1999: Part 1–2: General, structural fire design
- BS EN 1999: Part 1–3: Additional rules for structures susceptible to fatigue
- BS EN 1999: Part 1–4: Supplementary rules for trapezoidal sheeting
- BE EN 1999: Part 1–5: Supplementary rules for shell structures

Each part can be accompanied by a National Annex document. The National Annex states alternative procedures, values and recommendations where national choices may have to be made and provides national parameters and design values for the indicated clauses.

7.2 Symbols for structural design

The symbols used for structural design for equations and expression in many cases differ widely when comparing BS 8118 to Eurocode 9. A few of the main symbols common to both design codes are listed below, all other symbols that are used are listed and defined separately with associated equations and examples.

A = area
A_{eff} = effective area
A_{net} = net area
B = width(s)
C = constants
D = diameter
E = modulus of elasticity
G = shear modulus
I = second moment of area
I_{eff} = second moment of area for effective cross-section
i = radius of gyration
J = torsion constant
kN = kilonewton
L = length
L_{eff} = effective length
M = moment
mm = millimetre
m = metre
N = newton
n = number
R = radius of curvature

t = thickness
V = shear force
w = weld
α = coefficient of thermal expansion
β = slenderness parameter
γ = load factor(s)
δ = deflection
$\varepsilon = (\sqrt{250})/f_0$ [coefficient]
η = conversion/modifying factor
λ = slenderness parameter
μ = slip factor
v = Poisson's ratio
ρ = adjustment factor
σ = normal stress
τ = shear stress
ψ = parameter/stress ratio
x-x = axis along a member
y-y = axis of cross-section
z-z = axis of cross-section

The symbols that are used can vary in and for each design situation and may be different in other publications. However, the use of symbols should be kept consistent throughout the structural design and symbols should be labelled for ease of use and clarity for the reader of design documents.

7.3 Principle of member axes

It is important not to confuse the global axes X, Y, Z used in the coordinate system for modelling a structure in two or three dimensions, with the member axes, often called local axes. Member axes are aligned with the relevant member or element and may not line up with the global axes.

Figures 7.1 and 7.2 illustrate the local or member axes that are generally used in structural design. Designation of the axes is by small letters and often with an added accent as shown below.

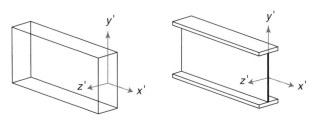

Figure 7.1 | Member axes

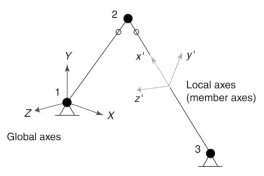

Figure 7.2 | Global and local axes

In most design codes the member axes are:

x-x (x') axis along a member
y-y (y') axis of cross-section (perpendicular to flanges)
z-z (z') axis of cross-section (parallel to flanges)

This notation is also used in Eurocode 9 (BS EN 1999).

7.4 Design basics

The principles of structural design of aluminium members are approached similarly by the British Standard (BS 8118 Part 1) and the new Eurocode 9 (EN 1999-1-1). In this chapter the methods and terms described in Eurocode 9 are used. The relevant design guidance to the British Standard is found in Clauses 4.3 and 4.4 of BS 8118: Part-1.

7.4.1 Classification of cross-section

Most standard structural members can be subdivided into flat plate elements. Thin plate elements can buckle and deform when subject to compressive stresses. This behaviour may possibly reduce the axial and flexural resistance well below the plastic moment resistance of a member. As this is independent of the length of a member it is described as local buckling. Local buckling is a critical factor in aluminium members. It is addressed in the classification of cross-sections, where a reduced capacity is advised for structural member design. To determine the reduced resistance of a member it must first be classified and then matched to one of the groups listed below.

- Class 1: Ductile. Class 1 cross-sections will be able to form plastic hinges with the rotational capacity required for plastic analysis, without any reduction in the resistance of the cross-section (see EN 1999-1-1, 6.1.4.2).

- Class 2: Compact. Class 2 cross-sections will be able to develop plastic moment resistance, but local buckling will result in limited rotational capacity only (see EN 1999-1-1, 6.1.4.2).
- Class 3: Semi-compact. Class 3 cross-sections will not be able to develop full plastic moment resistance due to local buckling, but the compression stress in the extreme fibre can reach to proof strength (see EN 1999-1-1, 6.1.4.2).
- Class 4: Slender. Class 4 cross-sections will not be able to reach proof strength i.e. local buckling will occur in one or more parts of the cross-section (see EN 1999-1-1, 6.1.4.2).

Compared to the above four groups specified in Eurocode 9, BS 8118 only uses three groups to classify aluminium cross-sections:

- Fully compact, where local buckling can be ignored.
- Semi-compact, where the section can develop full elastic moment resistance.
- Slender, where the member resistance is reduced by local buckling below the limiting stress of elastic bending.

The classification procedure involves firstly identifying the type of thin-walled part from given listed types aided by Figure 7.3:

- flat outstand parts
- flat internal parts
- curved internal parts

The next step is to determine the susceptibility to local buckling of a flat thin part by calculating its slenderness parameter β, from Equations (7.1)–(7.3). The equations listed below only apply to unreinforced parts:

$$\left.\begin{array}{l} \text{flat internal parts without stress gradient or} \\ \text{flat outstands without stress gradient or} \\ \text{peak compression at toe} \end{array}\right\} \quad \beta = b/t \quad (7.1)$$

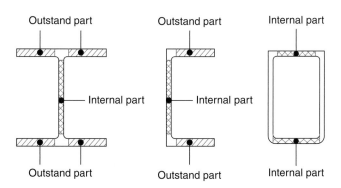

Figure 7.3 | Illustration of outstand and internal parts

internal parts with a stress gradient that results in a neutral axis at the centre $\left.\phantom{\begin{matrix}a\\b\end{matrix}}\right\}$ $\beta = 0.4 \times b/t$ (7.2)

internal parts with stress gradient and outstands with peak compression at root $\left.\phantom{\begin{matrix}a\\b\end{matrix}}\right\}$ $\beta = \eta \times b/t$ (7.3)

where

b = width of a cross-section part (or d)
t = thickness of a cross-section part
η = stress gradient factor given by:

$$\eta = 0.7 + (0.3 \times \psi) \text{ and } (1 \geq \psi \geq -1) \qquad (7.4)$$
$$\eta = 0.8/(1 - \psi) \text{ and } (\psi < -1) \qquad (7.5)$$

where ψ = ratio of stresses at the edges (see Figure 7.4)
Further information and gradient factors are given in EN 1999-1-1: 2007, Figure 6.2.

Where internal or outstanding parts are reinforced, three buckling modes need to be analysed and separate values of β need to be found for each mode. The buckling modes for reinforced parts are listed below:

- Mode 1: (a) Distortional buckling, where the reinforced part will buckle as one part and the reinforcement buckles with the same curvature as the part (see EN 1999-1-1, 6.1.4.3).
- Mode 2: (b) Individual buckling, where single parts (sub-parts) and the reinforcement buckle individually with the junction between them will be unaffected and straight (see EN 1999-1-1, 6.1.4.3).
- Mode 3: (c) Combination buckling, where single parts (sub-parts) buckles are superposed on the buckles of the entire part (see EN 1999-1-1, 6.1.4.3).

The three modes for local buckling of reinforced parts are illustrated in Figure 7.5.

Again, the next step is to determine the susceptibility to local buckling of a flat thin part by calculating its slenderness parameter β, from Equations (7.6)–(7.14). These equations only apply to reinforced parts and depend on the local buckling mode.

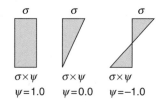

Figure 7.4 | Illustration for ratios of stresses (typical)

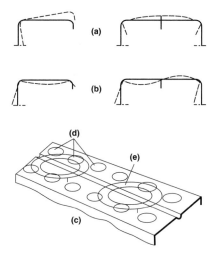

Figure 7.5 | Local buckling modes (Extracted from BS EN 1999-1-1, 6.3)

Values of β for mode 1

(a)   Uniform compression with standard reinforcement such as a single rib or lip of thickness equal to the part thickness *t*:

$$\beta = \eta \times b/t \tag{7.6}$$

where η is given by Equations (7.7)–(7.9):

(i)
$$\eta = 1 \Big/ \sqrt{1 + 0.1 \times (c/t - 1)^2} \tag{7.7}$$

for cases:

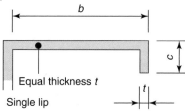

(ii)
$$\eta = 1 \Big/ \sqrt{1 + 2.5 \times \frac{(c/t - 1)^2}{(b/t)}} \geq 0.5 \tag{7.8}$$

for cases:

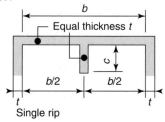

(iii)
$$\eta = 1 \Big/ \sqrt{1 + 4.5 \times \frac{(c/t - 1)^2}{(b/t)}} \geq 0.33 \qquad (7.9)$$

for cases:

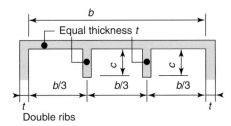

Double ribs

(b)   Uniform compression with non-standard reinforcement:

For any other single shape of reinforcement, the reinforcement is replaced by an equivalent rib or lip which is equal in thickness to the part. The value of c for the equivalent rib or lip is chosen so that the second moment of area of the reinforcement about the mid-plane of the plate part is equal to that of the non-standard reinforcement about the same plane. An alternative method is given in Eurocode 9, Clause 6.6 (EN 1999-1-1, 6.1.4.3).

(c)   Uniform compression with complex reinforcement:

Unusual shapes of reinforcement that are not suitable for the methods which have been described above:

$$\beta = (b/t) \times (\sigma_{cr0}/\sigma_{cr})^{0.4} \qquad (7.10)$$

where

σ_{cr0} = elastic critical stress for the unreinforced part (assuming simply supported edges of the parts)

σ_{cr} = elastic critical stress for the reinforced part (assuming simply supported edges of the parts)

(d)   Stress gradient:

$$\beta = (b/t) \times (\sigma_{cr0}/\sigma_{cr})^{0.4} \qquad (7.11)$$

where σ_{cr} and σ_{cr0} now relate to the stress at the more heavily compressed edge of the part.

Values of β for mode 2

(a)   The value(s) of the parameter β are determined separately for each sub-part as described above.

(b)   Uniform compression with shallow curved unreinforced internal part:

$$\beta = (b/t) \times 1 \Big/ \sqrt{1 + 0.006 \times \frac{b^4}{R^2 \times t^2}} \qquad (7.12)$$

where

> R = radius of curvature to the mid-thickness of material
> b = developed width of the part at mid-thickness of material
> t = thickness of the part

But this is only applicable if:

$$R/b > 0.1 \times b/t \tag{7.13}$$

(c) Thin-walled tube (compression and/or bending):

$$\beta = 3 \times \sqrt{D/t} \tag{7.14}$$

where D = diameter to mid-thickness of tube material and t = thickness of the part.

Classification of cross-sections
Using the above factors and parameters, parts of the cross-section and subsequently the entire cross-section can be classified i.e. associated with one of the four classes described earlier in this chapter.

Classification of the entire cross-section depends on the parts of the cross-section in compression, where individual parts of the cross-section can be placed in different classes. The cross-section is then classified on the basis of the grouping of the least favourable compression part.

However, the cross-section class is also directly dependent on the loading of the cross-section. It is possible that a cross-section belongs to a different class for axial compression, bending about the y-axis and bending about the z-axis, respectively. Cross-section classification is carried out separately for each action and not for combined actions. Figure 7.6 shows the grouping of parts in different classes for the same cross-section:

The classification of the cross-section is based on the slenderness parameter β as shown below (see EN 1999-1-1, 6.1.4.4).

- For parts in beam members:

$$\beta \le \beta_1 = \text{Class 1}$$
$$\beta_1 < \beta \le \beta_2 = \text{Class 2}$$
$$\beta_2 < \beta \le \beta_3 = \text{Class 3}$$
$$\beta_3 < \beta = \text{Class 4}$$

- For parts in strut members:

$$\beta \le \beta_2 = \text{Class 1 or Class 2}$$
$$\beta_2 < \beta \le \beta_3 = \text{Class 3}$$
$$\beta_3 < \beta = \text{Class 4}$$

The values for the parameters β_1, β_2 and β_3 can be obtained from Tables 7.1 and 7.2.

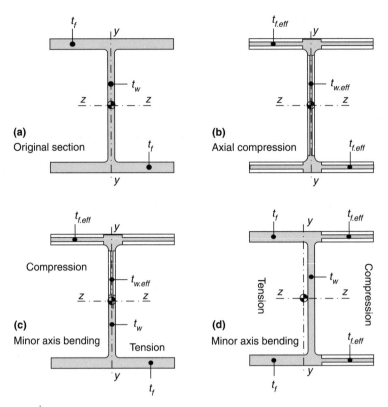

Figure 7.6 | Effective cross–section for classification

Table 7.1 | Slenderness parameters for internal parts ($\varepsilon = \sqrt{250/f_0}$ (N/mm²))

Buckling class	β_1 / ε	β_2 / ε	β_3 / ε
A (without welds)	11	16	22
A (with welds)	9	13	18
B (without welds)	13	16.5	18
B (with welds)	10	13.5	15

Eurocode 9, Table 3.2 gives the relevant buckling class (Class A or Class B) 'With welds' means a cross-section part having welding at an edge or within its width. For further guidance refer to EN 1999-1-1, Clause 6.1.4.4 and notes to EN 1999-1-1, Table 6.2

Table 7.2 | Slenderness parameters for outstand parts ($\varepsilon = \sqrt{250/f_0}$ (N/mm²))

Buckling class	β_1 / ε	β_2 / ε	β_3 / ε
A (without welds)	3	4.5	6
A (with welds)	2.5	4	5

Table 7.2 | (*Continued*)

Buckling class	β_1 / ε	β_2 / ε	β_3 / ε
B (without welds)	3.5	4.5	5
B (with welds)	3	3.5	4

Eurocode 9, Table 3.2 gives the relevant buckling class (Class A or Class B) 'With welds' means a cross-section part having welding at an edge or within its width. For further guidance refer to EN 1999-1-1, Clause 6.1.4.4 and notes to EN 1999-1-1, Table 6.2

Example 7.1: Classification of cross-section
A typical I-section taken from BS 1161: 1977

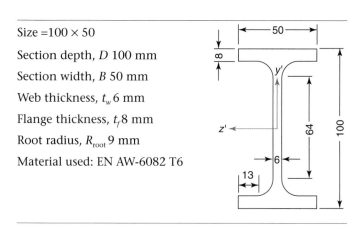

Size = 100×50

Section depth, D 100 mm

Section width, B 50 mm

Web thickness, t_w 6 mm

Flange thickness, t_f 8 mm

Root radius, R_{root} 9 mm

Material used: EN AW-6082 T6

For EN AW-6082 T6 EP/O, $f_0 = 260$ N/mm²

(EN 1999-1-1, Table 3.2b)

Buckling class = Class A
Bending about major axis (*y-y*)

(1) Classification of web part   (EN 1999-1-1, 6.1.4)

Internal cross-section part   (EN 1999-1-1, Figure 6.1)

$\beta_w = 0.4 \times b/t = 0.4 \times 64/6 = 4.27$
$\varepsilon_w = \sqrt{250/f_0} = \sqrt{250/260} = 0.981$
$\beta_{1.w} = 11 \times \varepsilon = 11 \times 0.981 = 10.791$ (EN 1999-1-1, Table 6.2)
$\beta_{2.w} = 16 \times \varepsilon = 16 \times 0.981 = 15.696$
$\beta_{3.w} = 22 \times \varepsilon = 22 \times 0.981 = 21.582$

Classification checks   (EN 1999-1-1, 6.1.4.4)

for Class 1: $\beta \leq \beta_1$ $4.27 \leq 10.791$ true
for Class 2: $\beta_1 < \beta \leq \beta_2$ $10.791 < 4.27 \leq 15.696$ false

for Class 3: $\beta_2 < \beta \le \beta_3$ $15.696 < 4.27 \le 21.582$ false
for Class 4: $\beta_3 < \beta$ $21.582 < 4.27$ false

Thus, the web part is assigned to Class 1.

(2)   Classification of flange parts   (EN 1999-1-1, 6.1.4)

Symmetrical outstand cross-section part   (EN 1999-1-1, Figure 6.1)

$\eta = 1.0$ (EN 1999-1-1, Figure 6.2)
$\beta_f = \eta \times b/t = 1 \times 13/8 = 1.63$
$\varepsilon_f = \sqrt{250/f_0} = \sqrt{250/260} = 0.981$
$\beta_{1.f} = 3.0 \times \varepsilon = 3.0 \times 0.981 = 2.943$ (EN 1999-1-1, Table 6.2)
$\beta_{2.f} = 4.5 \times \varepsilon = 4.5 \times 0.981 = 4.415$
$\beta_{3.f} = 6.0 \times \varepsilon = 6.0 \times 0.981 = 5.886$

Classification checks   (EN 1999-1-1, 6.1.4.4)

for Class 1: $\beta \le \beta_1$ $1.63 \le 2.943$ true
for Class 2: $\beta_1 < \beta \le \beta_2$ $2.943 < 1.63 \le 4.415$ false
for Class 3: $\beta_2 < \beta \le \beta_3$ $4.415 < 1.63 \le 5.886$ false
for Class 4: $\beta_3 < \beta$ $5.886 < 1.63$ false

Thus, the flange parts are assigned to Class 1.

(3)   Solution: Since the web and flange parts have been identified as Class 1 parts, the whole cross-section can be assigned to Class 1.

7.4.2   Local buckling

As discussed earlier, cross-section parts are limited in compression by local buckling. The resistance to compression of a cross-section part depends on the slenderness i.e. classification of the cross-section part. Therefore, local buckling of a single part, or more cross-section parts, can dominate the structural design of the member under consideration.

The effect on the resistance to compression of a cross-section part in relationship to its association with the classes laid out in Eurocode 9 is illustrated in Figure 7.7. Whilst the effect of local buckling for Classes 1–3 (non-slender parts) is negligible or small, the greater reduction in load capacity for Class 4 parts (slender parts) requires separate consideration.

The reduction of compression capacity due to local buckling is addressed by reducing the section area for slender parts. This method is applied in BS 8118: Part 1 and Eurocode BS EN 1999-1-1. For structural design to BS 8118: Part 1, the reduction factor (local buckling factor) k_L is obtained from Figure 4.5 in BS 8118: Part 1. A similar approach is used for designs to Eurocode 9 where the reduction factor or local buckling factor is given in Clause 6.1.5.

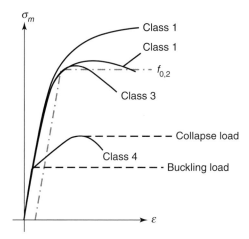

Figure 7.7 | Capacity dependency on slenderness of cross-section parts

For Eurocode 9 designs the factor ρ_c for slender parts (Class 4 parts) is calculated from:

$$\rho_c = \frac{C_1}{(\beta / \varepsilon)} - \frac{C_2}{(\beta / \varepsilon)^2} \tag{7.15}$$

For all other cross-section parts, where $\beta > \beta_3$, $\rho_c = 1.0$. For flat outstand parts in unsymmetrical cross-sections, ρ_c is given by Equation (7.15) or $\rho_c = 1.0$, but is limited to not greater than $120/(\beta/\varepsilon)^2$. The constants C_1 and C_2 are found in Table 6.3 of Eurocode 9 and in Table 7.3.

Table 7.3 | Constants C_1 and C_2 (see EN 1999–1-1, Table 6.3)

Buckling class	Internal parts		Outstand parts	
	C_1	C_2	C_1	C_2
A (without welds)	32	220	10	24
A (with welds)	29	198	9	20
B (without welds)	29	198	9	20
B (with welds)	25	150	8	16

Eurocode 9, Table 3.2 gives the relevant buckling class (Class A or Class B)
'With welds' means a cross-section part having welding at an edge or within its width. For further guidance refer to EN 1999-1-1, Clause 6.1.4.4 and notes to EN 1999-1-1, Table 6.2

All possible modes of buckling should be addressed for reinforced cross-section parts, and the lower value of ρ_c should be used for the subsequent design calculations. In the case of mode 1 buckling the factor ρ_c should be applied to the area of the reinforcement and to the plate thickness. The reduced sectional area of the slender part is then calculated by applying the local buckling factor ρ_c to the thickness t of the part.

Example 7.2: Effective area of cross-section with slender parts
Typical I-section

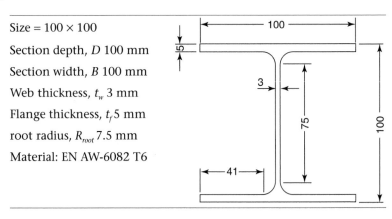

Size = 100 × 100

Section depth, D 100 mm

Section width, B 100 mm

Web thickness, t_w 3 mm

Flange thickness, t_f 5 mm

root radius, R_{root} 7.5 mm

Material: EN AW-6082 T6

For EN AW-6082 T6 EP/O, $f_0 = 250$ N/mm²

(EN 1999-1-1, Table 3.2b)

Buckling class = Class A, axial compression without bending.

(1) Web part   (EN 1999-1-1, 6.1.4/5)
       Internal cross-section part   (EN 1999-1-1, Figure 6.1)

β_w $= b/t = 75/3 = 25.00$
ε_w $= \sqrt{250/f_0} = \sqrt{250/250} = 1.00$
$\beta_{3.w}$ $= 22 \times \varepsilon = 22 \times 1.00 = 22.00$ (EN 1999-1-1, Table 6.2)
$\beta_3 < \beta$ 22.00 < 25.00, hence it is a slender cross-section part
β_w/ε $= 25.00/1.00 = 25$ (EN 1999-1-1, 6.1.5)
$(\beta_w/\varepsilon)^2 = 25^2 = 625$
$\rho_{c.w}$ $= 32/25 - 220/625 = 0.928$
$t_{w.eff}$ $= 3 \times 0.928 = 2.8$ mm

(2) Flange parts:
       Symmetrical outstand cross-section part   (EN 1999-1-1, Figure 6.1)

β_f $= b/t = 41/5 = 8.20$
ε_f $= \sqrt{250/f_0} = \sqrt{250/250} = 1.00$
$\beta_{3.f}$ $= 6 \times \varepsilon = 6 \times 1.00 = 6.00$ (EN 1999-1-1, Table 6.2)

$\beta_3 < \beta$ 6.00 < 8.20, hence it is a slender cross-section part

β_f/ε = 8.20/1.00 = 8.20   (EN 1999-1-1, 6.1.5)

$(\beta_f/\varepsilon)^2 = 8.20^2 = 67.24$

$\rho_{c.f}$ = 10/8.20 – 24/67.24 = 0.863

$t_{f.eff}$ = 5 × 0.863 = 4.3 mm

(3)   Solution:

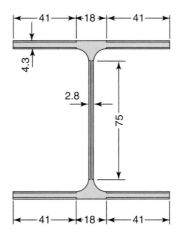

The above figure shows the reduced cross-section for structural design purposes. The properties of the original and reduced cross-sections are given in Table 7.4

Table 7.4 | Properties of original and reduced cross-section

Original section		Reduced section		Units
A_g	13.184	A_{eff}	11.886	cm²
I_y	253.145	$I_{y.eff}$	226.478	cm⁴
I_z	83.412	$I_{z.eff}$	71.810	cm⁴
I_t	1.121	$I_{t.eff}$	0.598	cm⁴
$W_{el.y}$	50.629	$W_{el.y.eff}$	45.296	cm³
$W_{el.z}$	16.682	$W_{el.z.eff}$	14.362	cm³
$W_{pl.y}$	55.670	$W_{pl.y.eff}$	49.936	cm³
$W_{pl.z}$	25.356	$W_{pl.z.eff}$	21.948	cm³

A = cross-sectional area
$I_{(y/z)}$ = moment of inertia
I_t = torsion constant
$W_{el.(y/z)}$ = elastic section modulus
$W_{pl.(y/z)}$ = plastic section modulus

7.4.3 Heat-affected zones softening

When welding aluminium alloy members, heat generated by the welding process reduces the properties of the material in the vicinity of the welds. The zones affected by this softening are known as heat-affected softening zones.

Both British Standard 8118 and Eurocode 9 give detailed guidance on how to make allowances for the loss of strength within heat-affected zones (HAZs). The reduction in strength can be severe and must not be overlooked. Only for parent material supplied in the annealed or T4 condition can HAZ softening be ignored.

The extent and severity of the heat-affected softening must be known for structural design. The severity of heat-affected softening is expressed by applying a reduction factor, where the strength properties of the parent material are reduced by the earlier mentioned reduction factor. It is also important to determine the extent of the heat-affected softening zones.

Extent of heat-affected zone
The extent of a HAZ is described by the distance b_{haz} which is measured from a weld as illustrated in Figure 7.8. The extent, i.e. dimension b_{haz}, is dependent on the type of welding (MIG or TIG), parent material and thickness of the parent material. Due to the greater heat input for a TIG weld, separate distances b_{haz} are given for material thickness of up to 6 mm. Table 7.5, taken from the new Eurocode 9, gives values

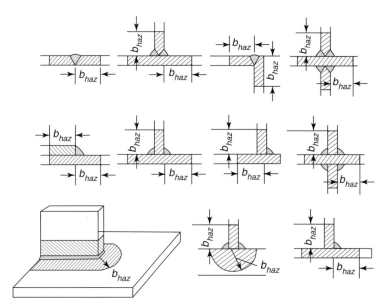

Figure 7.8 | Extend of heat–affected zones (Extracted from EN 1999–1–1, Figure 6.6)

Table 7.5 | Extent of HAZ depending on type of welding

Material thickness, *t* (mm)	MIG welding b_{haz} (mm)	TIG welding b_{haz} (mm)
$0 < t \leq 6$	20	30
$6 < t \leq 12$	30	30 (to 35)
$12 < t \leq 25$	35	35 (to 40)
$t < 25$	40	40 (to 50)

for b_{haz} in relation to the type of welding and material thickness. Good practice, which is to limit the effect of HAZ, can help to produce sections without reduction in the required primary strength.

Figure 7.9 illustrates the loss of strength at the vicinity of welds due to heat softening. Where the web plate is welded to the flange plates softening takes place in the flange and web part of the I-section. This will have an adverse consequence on the bending moment capacity of a beam. Welding T-sections forming an I-beam as shown in the second case in Figure 7.8 will retain the full bending moment capacity with only the web being weakened. This would subsequently reduce the shear resistance of the member. However, shear is often the non-dominant factor when designing aluminium beams. This example explains the importance of understanding the effect of the HAZ, especially when intending to use built-up welded sections, where the designer may be able to move the HAZs to less stressed locations, thus maximising the capacity of a structural member for its intended use.

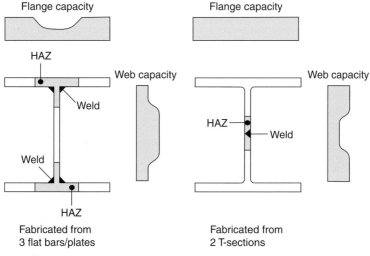

Figure 7.9 | Capacity loss due to HAZ

Severity of heat-affected zones

The severity of HAZs is expressed by the application of a reduction or HAZ factor. The material strength is then reduced by this reduction factor. This procedure is used in BS 8118 and Eurocode 9 (BS EN 1999). The reduction factors $\rho_{o.haz}$ and $\rho_{u.haz}$ can be calculated from Equations (7.16) and (7.17) as given in Clause 6.1.6.2 of EN 1999-1-1:

$$\rho_{o,haz} = f_{o,haz}/f_o \tag{7.16}$$

$$\rho_{u,haz} = f_{u,haz}/f_u \tag{7.17}$$

where

$\rho_{o.haz}$ = HAZ factor for 0.2% proof strength
$\rho_{u.haz}$ = HAZ factor for ultimate tensile strength
f_o = characteristic value of 0.2% proof strength
f_u = characteristic value of ultimate tensile strength
$f_{o.haz}$ = 0.2% proof strength in HAZ
$f_{u.haz}$ = ultimate tensile strength in HAZ

All of the above values can be readily found in Tables 3.2 of the Eurocode 9. To address the greater heat input for TIG welding, the values in that table or $\rho_{o.haz}$, $\rho_{u.haz}$, $f_{o.haz}$ and $f_{u.haz}$ are only to be used for MIG welding. A further reduction of the above HAZ values and factors is needed to allow for TIG welding. However, no reduction of the values and factors applies for material in temper O.

Table 7.6 gives the reduction factors for values based on MIG welding when TIG welding is used in welded aluminium joints. These reduction factors are applied to the values for $\rho_{o.haz}$, $\rho_{u.haz}$, $f_{o.haz}$ and $f_{u.haz}$ which are obtained from Tables 3.2 of Eurocode 9.

Table 7.6 | TIG reduction factors for HAZ

Material	Thickness (mm)	TIG reduction factor
3xxx 5xxx 8011A	≤6	1.0
3xxx 5xxx 8011A	>6	0.9
6xxx 7xxx	≤6	0.8
6xxx 7xxx	>6	0.8

At present no guidance is given for friction stir welding (FSW) in BS 8118 and the new Eurocode 9. Although less heat input is generally used for the welding process, the reduction factors for MIG welding can be used for calculating the HAZ strength of the welded material. The differences between FSW and MIG/TIG welding are illustrated in Figure 7.10.

Testing of friction stir welded joints in AW6082-T6 carried out by Adamowski and Szkodo (2007) resulting in the joints failing within the HAZ. Thus the reduction in the strength of the material can be assumed to be similar to that for fusion welding (MIG/TIG).

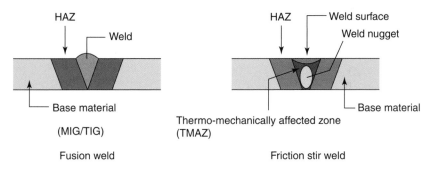

Figure 7.10 | Types of welding showing HAZ

7.5 Tension members

Tension members (ties) i.e. members subject to longitudinal tensile load are perhaps the easiest to design. When a tensile load is applied to the longitudinal axis at the centroid of a cross-section the member is classified as axially loaded. The stress generated in the material by this tensile axial load is known as the tensile stress. Tensile stress is assumed to be uniform over the entire cross-section. However, this is limited to a straight member. With no reduction of the cross-sectional area due to holes and HAZs the design check is a simple process of checking the tensile capacity of the cross-section from the following basic equation:

$$\frac{Tensile\ Force}{Area} \leq Design\ Value \tag{7.18}$$

where the tensile force = applied (factored) tensile load;
area = cross-sectional area;
and the design value = allowable (design) stress.

Despite the simple approach for the structural design of tension members, consideration and allowances must be made for reduced cross-sectional area, heat-affected localised capacity reduction, combined action such as axial tension and bending, and also the event of load reversal changing the member stress to compression.

The design criteria for tension members can be summarised as:

- member axially loaded
- reduced cross-sectional area (holes or HAZ)
- combined action (tension and bending)
- possibility of stress reversal

Failure of tension members occurs due to ductile yield i.e. excessive deformation (elongation) or fracture at the reduced cross-section. Excessive deformation is caused by section yielding along the length of the member. Fracture can occur if the stress at the net area of the cross-section reaches or surpasses the ultimate stress of the material.

The design methods, described by BS 8118: Part 1: Clause 4.6 and Eurocode 9: Clause 6.2.3 for calculating the design resistance to tensile forces, are the same. The factored axial tensile loading should not exceed the design tension resistance of the section. The appropriate equation from Eurocode 9, which is representative of both design codes, is:

$$\frac{N_{Ed}}{N_{t,Rd}} \leq 1.0 \qquad (7.19)$$

where N_{Ed} = design normal force and $N_{t,Rd}$ = design tensile resistance.

Considering the three possible failure modes, general (ductile) yielding, fracture at reduced cross-section due holes and failure at a section with HAZ, the design tensile resistance $N_{t,Rd}$ is given by the lesser of the design resistance to general yielding $N_{o,Rd}$ and the design resistance to axial force of a net cross-section with holes or effected by HAZ $N_{u,Rd}$. The relevant equations, taken from Clause 6.2.3 of Eurocode 9, are listed below:

- for general yielding:

$$N_{o,Rd} = A_g \times f_o / \gamma_{M1} \qquad (7.20)$$

- for local failure of a section with holes:

$$N_{u,Rd} = 0.9 \times A_{net} \times f_u / \gamma_{M2} \qquad (7.21)$$

- for local failure of a section with HAZ:

$$N_{u,Rd} = A_{eff} \times f_u / \gamma_{M2} \qquad (7.22)$$

where

A_{eff} = effective area of cross-section
A_g = gross area of cross-section
A_{net} = net area of cross-section
f_o = characteristic value of 0.2% proof strength
f_u = characteristic value ultimate tensile strength
γ_{M1} = partial factor for resistance of cross-section
γ_{M2} = partial factor for tensile fracture

Example 7.3: Tension resistance for general yielding, failure with holes and HAZ

Check the tensile resistance of a flat bar 10×60 mm, for EN AW-6082 T6 ER/B.

The data for the material is: (EN 1999-1-1, Table 3.2b)

$f_o = 250$ N/mm²
$f_u = 295$ N/mm²
$f_{o.haz} = 125$ N/mm²
$f_{u.haz} = 185$ N/mm²
$\rho_{o.haz} = 0.50$
$\rho_{u.haz} = 0.63$
$A_g = 600$ mm²

Partial safety factors   (EN 1999-1-1, 6.1.3)

$\gamma_{M1} = 1.10$
$\gamma_{M2} = 1.25$

(1) General yielding of cross-section   (EN 1999-1-1, 6.2.3 (6.18))

$$N_{o,Rd} = A_g \times f_o / \gamma_{M1}$$

$$600 \times 250/1.10 = \mathbf{136\,364\ N}$$

(2) Fracture/failure of cross-section with hole(s)   (EN 1999-1-1, 6.2.3 (6.19a))

$$N_{u,Rd} = 0.9 \times A_{net} \times f_u / \gamma_{M2}$$

$$A_{net} = 600 - (13 \times 10) = 470 \text{ mm}^2 \ \textit{(for ø13 mm hole)}$$

$$N_{u,Rd} = 0.9 \times 470 \times 295/1.25 = \mathbf{99\,828\ N}$$

(3) Failure of cross-section with HAZ   (EN 1999-1-1, 6.2.3 (6.19b))

$$N_{u,Rd} = A_{eff} \times f_u / \gamma_{M2}$$

$$A_{eff} = 0.63 \times 600 = 378 \text{ mm}^2$$

$$N_{u,Rd} = 378 \times 295/1.25 = \mathbf{89\,208\ N}$$

(4) Solution: The tie 10×60 mm, EN AW-6082 T6 ER/B, has the lowest tension resistance in the welded condition. A maximal factored tensile force of 89.208 kN can be applied to the member. Bolting would improve the member's tension resistance by approximately 11%.

7.6   Compression members

Compression members (struts) are members/elements subject to longitudinal compressive load. When a compressive load is applied to the longitudinal axis at the centroid of a cross-section the member is classified as axially loaded. The stress generated in the material by this compressive axial load is known as the compressive stress. Compressive stress is assumed to be uniform over the entire cross-section.

Differing from the simple structural design of tension members, the design of compression members requires further considerations associated with additional failure modes depending on the section class and length (slenderness) of the strut. Nevertheless, some factors are the same. The design criteria for compression members can be summarised as:

- member is axially loaded
- reduced cross-sectional area (holes or HAZ)
- combined action (compression and bending)
- length of strut
- section class

Compression members fail due to local buckling, member buckling and compression yielding of the cross-section. Consequently, these design checks must be conducted to ensure the structural integrity of the compression member.

7.6.1   Local buckling

The effect of local buckling on a member's compression capacity has been discussed in Sections 7.4.1 and 7.4.2. Buckling of cross-sectional parts is accounted for by the section classification in the four cross-section classes, Classes 1–4, and reduction of the cross-sectional area for slender Class 4 cross-sections.

The method for reducing the cross-sectional area, i.e. determination of the effective area A_{eff}, is described in Section 7.4.2. The value of the effective cross-section area is then used in the calculations to compute the design resistance to axial compression (local squashing) and member buckling (flexural and torsional).

7.6.2   Compressive yielding

Compressive yielding of the cross-section is often called local squashing. It describes the failure of the cross-section to support the axial thrust. For columns the critical location of local squashing is usually the base. For struts the least favourable cross-sectional area must be checked for local squashing.

Similarly to the method for designing tension members, the factored design value of the axial compression should not exceed the design compressive resistance of the section. This is illustrated in the equation cited from Eurocode 9 and shown below:

$$\frac{N_{Ed}}{N_{c,Rd}} \leq 1.0 \tag{7.23}$$

where N_{Ed} = design normal force and $N_{c,Rd}$ = design compression resistance.

Two failure modes are considered by the methods of Eurocode 9: first, the effective cross-section with deductions for unfilled holes and HAZ; and secondly, the effective cross-section with deductions for local buckling and HAZ. Therefore, the design resistance for uniform compression $N_{c,Rd}$ is given by the two equations taken from Eurocode 9 Clause 6.2.4. The resistance for compressive yielding should be checked along the member and taken at the least favourable section of the member. Again, Equations (7.24) and (7.25) are taken from Eurocode 9, however, similar expressions can be found in BS 8118: Part 1: Clause 4.7.7 Local squashing.

The design resistance for uniform compression $N_{c,Rd}$ should be taken as the lesser of $N_{u,Rd}$ and $N_{c,Rd}$, obtained from the equations given in Clause 6.2.4.

• For sections with unfilled holes:

$$N_{u,Rd} = A_{net} \times f_u / \gamma_{M2} \tag{7.24}$$

• For other sections:

$$N_{c,R\,d} = A_{eff} \times f_o / \gamma_{M1} \tag{7.25}$$

where

A_{eff} = effective area of cross-section
A_{net} = net area of cross-section
γ_{M1} = partial factor for resistance of cross-section
γ_{M2} = partial factor for tensile fracture

7.6.3 Member buckling

Unlike the local buckling of member parts, the overall buckling of a member is dependent on its effective length. Long slender struts will fail at a much lower load by elastic member buckling or lateral bending. Struts are therefore divided into three groups:

1) short struts
2) medium-length struts
3) slender struts

As discussed earlier, long and slender struts will fail by member buckling, short struts will usually fail by compressive yielding, and struts in between these two groups (medium struts) can fail by a combination of local squashing and member buckling.

The maximum axial load a strut can support is controlled by the material that has been used, the cross-sectional properties, and also the slenderness of the strut. The slenderness of a structural member is defined as its effective length divided by the radius of gyration taken about the appropriate axis:

$$\lambda = L_{eff}/i \qquad (7.26)$$

where

λ = slenderness

L_{eff} = effective length

i = radius of gyration of gross cross-section

The effective length or buckling length L_{cr} is found by applying the buckling length factor k to the strut length. This method as described in the Eurocode 9 is the same in BS 8118 with matching buckling length factors k in both design codes. The buckling length factor k takes into account the strut end or support conditions as the rigidity at the supports influences the resistance against member buckling. The buckling length factors given below are based on the research by Leonhard Euler (1707–1783). However, these factors have been altered to allow for shear and reduced stiffness of connections and therefore give a longer effective length than that obtained by Euler's theory.

It is important to provide the restraint at the supports i.e. connections as shown in Table 7.7, but also sufficient rotational stiffness to give the support condition used for consequent calculations. The buckling length L_{cr} is obtained from:

$$L_{cr} = L \times k \qquad (7.27)$$

where L = strut length and k = buckling length factor (from Table 7.7).

7.6.3.1 *Flexural buckling*

If a straight column or strut is loaded axially with a load greater than the critical load N_{cr}, flexural buckling will occur. Hence, the strut will deflect considerably sideways causing failure due to member buckling. Flexural buckling occurs about the axis with the largest slenderness ratio, and the smallest radius of gyration.

The symbolic representation in the right-hand column in Figure 7.11 illustrates the flexural buckling failure of an overloaded strut. The buckling analysis was carried out with the aid of advanced

Table 7.7 | Buckling length factors k to British and European Codes

End conditions	k	Symbolic
(No. 1) Held in position and restrained in direction at both ends	0.70	
(No. 2) Held in position at both ends and restrained in direction at one end	0.85	
(No. 3) Held in position at both ends, but not restrained in direction	1.00	
(No. 4) Held in position at one end, and restrained in direction at both ends	1.25	
(No. 5) Held in position and restrained in direction at one end, and partially restrained in direction but not held in position at the other end	1.50	
(No. 6) Held in position and restrained in direction at one end, but not held in position or restraint at the other end	2.00	

Figure 7.11 | Typical flexural buckling of a strut

three-dimensional analysis software and the resulting plot shows the expected curvature of the column.

The column was pinned at bottom and top with the load applied axially at the top node (end condition No. 3 from Table 7.7). With no other load acting on the strut, the buckling analysis shows the displacement buckling about the weaker axis of the universal beam (UB) type column as described above. Reduction of the applied load to less than the critical buckling load produces a plot with a straight column with negligible sideways displacement.

To account for slenderness, imperfections, welding and local buckling, the equation to obtain the design resistance to axial compression is extended with two reduction factors added to the initial formula. The two reduction factors for the Eurocode 9 approach are:

- reduction for the relevant buckling mode, χ
- reduction for weakening effects of welding, κ

The reduction factors for flexural buckling are dependent on the material type i.e. material buckling class and the relative slenderness but also the area of HAZ. Values and guidance to determine the appropriate reduction factors is given in Clause 6.3.1.2 of Eurocode 9.

The relative slenderness is given by the square root of the ratio of design resistance to compressive yielding and the elastic critical force for flexural buckling:

$$\bar{\lambda} = \sqrt{\frac{A_{eff} \times f_0}{N_{cr}}} = \frac{L_{cr}}{i \times \pi} \sqrt{\frac{A_{eff} \times f_0}{A_g \times E}} \qquad (7.28)$$

where

L_{cr} = buckling length in buckling pane considered
i = radius of gyration (gross cross-section)
A_{eff} = effective area allowing for local buckling
A_g = gross area of cross-section
E = modulus of elasticity

The reduction factor for flexural buckling χ can be obtained direct from Figure 7.12 extracted from Eurocode 9 where the arrows indicate an arbitrary example. However, the value of the reduction factor χ for flexural buckling can be calculated by the methods given in Clause 6.3.1.2 of Eurocode 9.

The reduction factor κ allowing for the weakening effects of welding can be compiled by equations and guidance given in Clauses 6.3.1.2 (BS EN 1999-1-1, Table 6.5) and 6.3.3.3. The procedure of Clause 6.3.1.2, shown below, is only applicable for longitudinal welds.

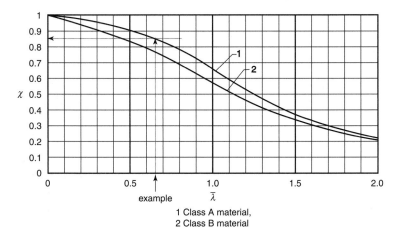

1 Class A material,
2 Class B material

Figure 7.12 | Reduction factor χ for flexural buckling (Extracted from EN 1999-1-1, Figure 6.11)

- Material for which buckling Class 'A' applies:

$$\kappa = 1 - \left(1 - \frac{A_1}{A}\right) \times 10^{-\bar{\lambda}} - \left(0.05 + 0.1 \times \frac{A_1}{A}\right) \times \bar{\lambda}^{-1.3(1-\bar{\lambda})} \qquad (7.29)$$

where: $A_1 = A - A_{haz} \times (1 - \rho_{o,haz})$ and A_{haz} = area of HAZ.

- Material for which buckling class 'B' applies:

  for $\bar{\lambda} \leq 0.2$: $\quad \kappa = 1.0$

  for $\bar{\lambda} > 0.2$: $\quad \kappa = 1 + 0.04 \times \left(4 \times \bar{\lambda}\right)^{(0.5-\bar{\lambda})} - 0.22 \times \bar{\lambda}^{-1.4(1-\bar{\lambda})} \qquad (7.30)$

For transverse welding the reduction factor κ can be found from the equation in Clause 6.3.3.3:

$$k = \omega_x = \frac{\rho_{u.haz} \times f_u / \gamma_{M2}}{f_0 / \gamma_{M1}} \qquad (7.31)$$

where

$\rho_{u,haz}$ = reduction factor for HAZ
f_o = characteristic value of 0.2% proof strength
f_u = characteristic value ultimate tensile strength
γ_{M1} = partial factor for resistance of cross-section
γ_{M2} = partial factor for tensile fracture

The design buckling resistance against flexural buckling $N_{b.f,Rd}$ of a compression member is given by:

$$N_{b.f,Rd} = \kappa \times \chi \times A_{eff} / \gamma_{M1} \qquad (7.32)$$

where κ = reduction factor – weakening of welding and χ = reduction factor for flexural buckling.

7.6.3.2 Torsional buckling and torsional-flexural buckling

Whilst flexural buckling can occur in any compression member if the member falls into one of the groups of medium-length struts or slender struts, torsional buckling mostly occurs in open cross-sections with slender cross-sectional parts, where the member is twisted about its longitudinal axis.

Torsional-flexural buckling takes place in struts that have an asymmetrical cross-section such as channels, tees and angle sections (see Figure 7.13). Here a combination of longitudinal bending and twisting about the longitudinal axis occurs. This phenomenon is based on combined twisting about the shear centre plus a translation of the shear centre.

Table 7.8 summarises the most likely buckling/failure mode of axially loaded compression members. Both design codes, BS 8118 and Eurocode 9, consider the susceptibility of cross-sections to the three different buckling modes and allow ignoring the possibility of torsional and torsional-flexural buckling for the following cross-section types:

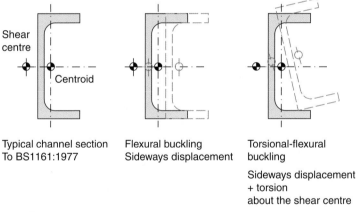

Typical channel section
To BS1161:1977

Flexural buckling
Sideways displacement

Torsional-flexural buckling

Sideways displacement
+ torsion
about the shear centre

Figure 7.13 | Principle of torsional–flexural buckling

Table 7.8 | Possible buckling modes for cross–section types

Cross-section type	Buckling mode		
	Flexural	**Torsional**	**Torsional-flexural**
Closed	✓		
Open, doubly symmetrical	✓	✓	
Open, singly symmetrical	✓		✓
Open, unsymmetrical			✓

- hollow sections (closed sections)
- doubly symmetrical I-sections
- sections composed entirely of radiating outstands (angles, tees etc.)
  if classified as Class 1 or Class 2

The buckling resistance to torsional and torsional-flexural buckling is calculated following the same method as applied for the determination of the resistance to flexural buckling and is given by the same equation, however, the reduction factor χ for torsional and torsion-flexural buckling to be calculated following the guidance given in Clause 6.3.1.2 of Eurocode 9 using the relevant parts for torsional and torsional-flexural buckling.

As before the relative slenderness is given by the square root of the ratio of design resistance to compressive yielding and the elastic critical force for torsional buckling:

$$\bar{\lambda} = \sqrt{\frac{A_{eff} \times f_o}{N_{cr}}} \tag{7.33}$$

where

A_{eff} = effective area for torsional or torsional-flexural buckling
f_o = characteristic value of 0.2% proof strength
N_{cr} = elastic critical load for torsional buckling

The procedure for finding the elastic critical load for torsional and torsional-flexural buckling is laborious and it is recommended to utilise cross-sections with good resistance to torsion and exempt from the requirement to check for failure due torsional and torsional-flexural buckling. Nonetheless, the equation to compute the elastic critical load as given in the Eurocode 9 (Annex I.3) is listed below.

Citing from the Annex I.3 of Eurocode 9 (EN 1999–1-1:2007): the elastic critical axial force N_{cr} for torsional and torsional-flexural buckling of a member if uniform cross-section, under stated condition of restraint at each end and subject to uniform axial force in the centre of gravity is given by:

$$(N_{cr,y} - N_{cr})(N_{cr,z} - N_{cr})(N_{cr,T} - N_{cr})i_s^2$$
$$-\alpha_{zw}z_s^2 N_{cr}^2(N_{cr,y} - N_{cr}) - \alpha_{yw}y_s^2 N_{cr}^2(N_{cr,z} - N_{cr}) = 0 \tag{7.34}$$

The reduction factor for torsional and torsional-flexural buckling χ can then be obtained directly from Figure 7.14 extracted from Eurocode 9 where the arrows indicate an arbitrary example. Again, the value of the

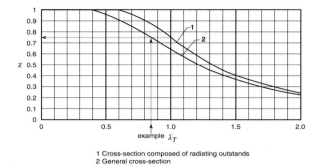

1 Cross-section composed of radiating outstands
2 General cross-section

Figure 7.14 | Reduction factor χ for torsional and torsional-flexural buckling (Extracted from EN 1999–1–1, Figure 6.12)

reduction factor χ for torsional and torsional-flexural buckling can be calculated by the means given in Clause 6.3.1.2 of Eurocode 9.

The method used to obtain the reduction factor to allow for weakening due to welding is the same as the one described in Section 7.6.3.1 of this chapter and similar factors and coefficients apply.

The design buckling resistance against torsional and/or torsional-flexural buckling $N_{b.t,Rd}$ of a compression member is given by:

$$N_{b.t,rd} = \kappa \times \chi \times A_{eff}/\gamma_{m1} \tag{7.35}$$

where κ = reduction factor – weakening of welding;
and χ = reduction factor for torsional and torsional-flexural buckling.

Example 7.4: Resistance to axial compression
Check the compression resistance of the I-section used in Example 7.1. There is no welding or unfilled holes along the section.

Size = 100×50	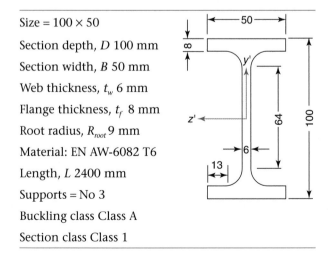
Section depth, D 100 mm	
Section width, B 50 mm	
Web thickness, t_w 6 mm	
Flange thickness, t_f 8 mm	
Root radius, R_{root} 9 mm	
Material: EN AW-6082 T6	
Length, L 2400 mm	
Supports = No 3	
Buckling class Class A	
Section class Class 1	

Material data   (EN 1999-1-1, Table 3.2b)

For EN AW-6082 T6 EP/O:

$$f_o = 260 \text{ N/mm}^2$$

$$f_u = 310 \text{ N/mm}^2$$

Partial safety factors   (EN 1999-1-1, 6.1.3)

$$\gamma_{M1} = 1.10$$

$$\gamma_{M2} = 1.25$$

(1) General yielding of cross-section   (EN 1999-1-1, 6.2.4 (6.22))

$$N_{c,Rd} = A_{eff} \times f_o/\gamma_{M1} = 1373 \times 260/1.10 = \mathbf{324\,527 \text{ N}}$$

(2) Flexural buckling:

$$\kappa = 1.00 \qquad \text{(for extruded section without weld)}$$

$$L_{cr} = L \times k = 2400 \times 1.00 = \mathbf{2400 \text{ mm}}$$

                                    (EN 1999-1-1, 6.3.1.3)

$$i_y = 39 \text{ mm}$$

$$i_z = 11 \text{ mm}$$

$$E = 70\,000 \text{ N/mm}^2 \qquad \text{(EN 1999-1-1, 3.2.5)}$$

$$\bar{\lambda} = \sqrt{\frac{A_{eff} \times f_o}{N_{cr}}} = \frac{L_{cr}}{i \times \pi}\sqrt{\frac{A_{eff} \times f_o}{A_g \times E}} = (2400/11\pi)\surd(260/70\,000) = \mathbf{4.23}$$

                                    (EN 1999-1-1, 6.3.1.2(6.51))

$$\bar{\lambda}_o = 0.10$$
$$\alpha = 0.20 \qquad \text{(EN 1999-1-1, Table 6.6)}$$

$$\phi = 0.5 \times \left(1 + \alpha\left(\bar{\lambda} - \bar{\lambda}_0\right) + \bar{\lambda}^2\right) = 0.5(1 + 0.2(4.23 - 0.1) + 4.23^2) = \mathbf{9.859}$$

                                    (EN 1999-1-1, 6.3.1.2(6.51))

$$x = \frac{1}{\phi + \sqrt{\phi^2 - \bar{\lambda}^2}} \text{ but } \chi < 1.00 \qquad \text{(EN 1999-1-1, 6.3.1.2(6.50))}$$

$$\chi = \mathbf{0.05}$$
$$N_{b,f,Rd} = \kappa \times \chi \times A_{eff}/\gamma_{M1} = 1.0 \times 0.05 \times 1373 \times 260/1.10 = \mathbf{16\,226 \text{ N}}$$
                                    (EN 1999-1-1, 6.3.1.1(6.49))

(3) Torsional and torsional-flexural buckling: No design checks are required for open doubly symmetrical cross-sections. (see EN 1999-1-1, 6.3.1.4(Note))

(4) Solution: The 100 × 50 I-section with a length of 2,400 mm would be able to support 16,226 N or 16.226 kN applied axial compression.

7.7 Flexural members

Flexural members are probably the most commonly used structural elements. Often called beams or beam-columns, they are loaded laterally, uniaxially or biaxially. They are regularly in combination with axial compression or tension, regularly seen with beam-columns or within frameworks. Whilst beams mainly carry vertical gravitational loads and usually span horizontally between two or more supports, beam-columns or beam-struts carry horizontal loads such as wind loads and span vertically between their supports. Typical definitions are shown in Figure 7.15.

As the aluminium alloys have a yield strength similar to that of mild steel, it might seem opportune to use the same design criteria and approach as for designing steel beams. However, with some of the unique characteristics of aluminium alloys, the structural design of an aluminium alloy beam is dissimilar in some aspects to the design of a steel beam (see Table 7.9).

The main difference from designing steel beams is the lower modulus of elasticity of aluminium members (70 000 N/mm²) compared to mild steel (210 000 N/mm²). Thus the aluminium alloy beam will deflect three times as much as the same cross-section made from mild steel. Therefore, displacement might be the governing design criteria for an aluminium beam. However, the lower modulus of elasticity will also affect the resistance to local buckling of the compression parts of an aluminium alloy cross-section, giving a lesser resistance to the buckling of compression cross-section parts compared to mild steel cross-sections.

As discussed in the earlier chapters of this book, local weakening due to welding requires additional design considerations when using aluminium alloy beams instead of mild steel beams. To account for the

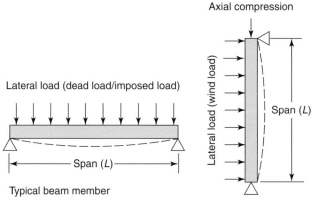

Figure 7.15 Typical flexural members

Table 7.9 | Material strength to Eurocodes EC3 and EC9

	Mild steel Grade S275	**Aluminium alloy 6082-T6**
Yield strength	275 N/mm²	250 N/mm²
Ultimate tensile strength	360 N/mm²	290 N/mm²

effects of local buckling and HAZ, the shape factor α is introduced in the Eurocode 9 design method. The typical shape factor α is the ratio of the plastic modulus W_{pl} to the elastic modulus W_{el} of the section, to make allowances for local buckling and welding. The relevant shape factors for cross-section classes and weld-free and welded section are given in Table 6.4 of BS EN 1999-1-1.

As examined in Sections 7.4.2 and 7.4.3 earlier in this chapter, the effective area of a cross-section or cross-section parts is calculated by reducing the part thickness accordingly. This method is utilised to determine the effective section moduli for Class 4 section or heat-affected sections (see Figure 7.16). Table 7.10 gives values for reduced part thickness for parts of cross-sections affected by local buckling and/or weakening by welding (HAZ).

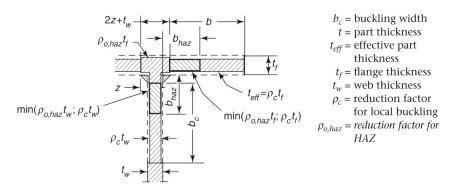

b_c = buckling width
t = part thickness
t_{eff} = effective part thickness
t_f = flange thickness
t_w = web thickness
ρ_c = reduction factor for local buckling
$\rho_{o,haz}$ = *reduction factor for HAZ*

Figure 7.16 | Effective thickness of Class 4 cross-section parts with HAZ (Extracted from EN 1999-1-1, Figure 6.9)

Table 7.10 | Values for reduced part thickness t_{eff} (see Figure 7.16 for key to symbols)

Cross-section class	**Without HAZ**	**With HAZ**
1	$t_{eff} = t$	$t_{eff} = \rho_{o,haz} \times t$
2	$t_{eff} = t$	$t_{eff} = \rho_{o,haz} \times t$
3	$t_{eff} = t$	$t_{eff} = \rho_{o,haz} \times t$
4	$t_{eff} = \rho_c \times t$	$t_{eff} = \min(\rho_c \times t; \rho_{o,haz} \times t)$

7.7.1 Bending only members

Bending members are structural elements subjected to loads that are usually applied perpendicular to their longitudinal axis. Bending members are often installed horizontally and loaded on the narrow/top face. This produces bending about the major axis of the beam and is generally the most efficient configuration for resisting bending actions. The failure modes for flexural members in pure bending are:

- bending moment
- shear
- web bearing
- lateral-torsional buckling (LTB)

7.7.1.1 Uniaxial bending to Eurocode 9

The resistance to bending of a beam can be found from beam bending theory, resulting in stress = moment divided by the section modulus, $\sigma = M/W$. Considering the weakening due to local buckling and welding, the design resistance M_{Rd} for bending about one principal axis of a cross-section is given by the lesser of the design resistance for bending of the net cross-section at holes $M_{u,Rd}$ and of the cross-section with HAZ and/or local buckling $M_{c,Rd}$:

$$M_{u,Rd} = W_{net} \times f_u / \gamma_{M2} \tag{7.36}$$

$$M_{c,Rd} = \alpha \times W_{el} \times f_o / \gamma_{M1} \tag{7.37}$$

where

α = shape factor
f_o = characteristic value of 0.2% proof strength
f_u = characteristic value ultimate tensile strength
γ_{M1} = partial factor for resistance of cross-section
γ_{M2} = partial factor for tensile fracture
W_{el} = elastic modulus of gross section
W_{net} = elastic modulus of net section

In addition, it should be noted that W_{net} is the elastic modulus of the net section allowing for holes and HAZ softening, if welded. The latter deduction is based on the reduced thickness of $\rho_{u,haz} \times t$ (EN 1999-1-1, 6.2.5.1).

7.7.1.2 Biaxial bending to Eurocode 9

When a member is subjected to moments about both principal axes, biaxial bending results and stresses introduced by biaxial bending must be checked. The condition for pure bending members is expressed in BS 8118: Part 1 adding the utilisation ratios of each axis to give less or equal to 1 = 100%:

$$\frac{M_{y,Ed}}{M_{y,Rd}} + \frac{M_{z,Ed}}{M_{z,Rd}} \leq 1.00 \qquad (7.38)$$

where

$M_{y,Ed}$ = design moment about y-axis
$M_{z,Ed}$ = design moment about z-axis
$M_{y,Rd}$ = moment resistance about y-axis
$M_{z,Rd}$ = moment resistance about z-axis

However, the method described in the Eurocode 9 (EN 1999-1-1) is more defined and additional factors are added to determine the adequacy of a member to resist biaxial bending. The factors used to check the structural sufficiency of a member in biaxial bending account for localised welds (ω_0) and LTB (χ_{LT}, ω_{xLT}). Taking a conservative approach, a section without welds and restrained to prevent LTB can be checked using the same equation as applied in BS 8118 and noted earlier. For non-slender cross-sections without welds it might be beneficial to use the more detailed calculation making use of the exponents γ_0 and ξ_0, both based on the shape factor α and specified in EN 1999-1-1, Clause 6.2.9.1.

$\gamma_0 = 1.0$ or may alternatively taken as α_z^2 but $1 \leq \gamma_0 \leq 1.56$ (7.39)

$\xi_0 = 1.0$ or may alternatively taken as α_z^2 but $1 \leq \xi_0 \leq 1.56$ (7.40)

where γ_0 = exponent and ξ_0 = exponent.

For members affected by HAZ softening, two conditions will give different factors ω_0:

1) For sections with HAZ softening with a specified location along the length and if the softening does *not* extend longitudinally a distance greater than the least width of the member ω_0 is obtained by:

$$\omega_0 = (\rho_{u,haz} \times f_u / \gamma_{M2})/(f_o / \gamma_{M1}) \qquad (7.41)$$

2) For sections with HAZ softening extending longitudinally a distance greater than the least width of the member ω_0 is obtained as:

$$\omega_0 = \rho_{o,haz} \qquad (7.42)$$

For members in biaxial bending the resistance to bending of both member axes must be checked as described earlier. Taking each axis in term the design values of the resistance to bending can be found from:

$$M_{y,Rd} = \alpha_y \times W_{y,el} \times f_o / \gamma_{M1} \qquad (7.43)$$

$$M_{z,Rd} = \alpha_z \times W_{z,el} \times f_o / \gamma_{M1} \qquad (7.44)$$

where

$M_{y,Rd}$ = moment resistance about y-axis
$M_{z,Rd}$ = moment resistance about z-axis
α_y = shape factor bending about y-axis
α_z = shape factor bending about z-axis
$W_{y,el}$ = elastic section modulus about y-axis
$W_{z,el}$ = elastic section modulus about z-axis
f_o = characteristic value of 0.2% proof strength
γ_{M1} = partial factor for resistance of cross-section

With the single axis moment resistance established, the combined bending is considered for both axes acting simultaneously. And the added utilisation ratios for both axes should not be greater than 1.0 = 100%. In compliance with EN 1999-1-1 the total utilisation for biaxial bending should be calculated from the following equations:

- For open cross-sections:

$$\left(\frac{M_{y,ED}}{\omega_0 \times M_{y,Rd}}\right)^{Y_0} + \left(\frac{M_{z,Ed}}{\omega_0 \times M_{z,Rd}}\right)^{\xi_0} \leq 1.00 \qquad (7.45)$$

- For hollow section and solid cross-sections:

$$\left[\left(\frac{M_{y.Ed}}{\omega_0 \times M_{y,Rd}}\right)^{1.7} + \left(\frac{M_{z,Rd}}{\omega_0 \times M_{z,Rd}}\right)^{1.7}\right]^{0.6} \leq 1.00 \qquad (7.46)$$

7.7.1.3   Shear to Eurocode 9

Checking the shear force resistance of a member involves two essential checks: first, for shear yielding; and secondly, for shear buckling of the web parts. Both design codes (BS 8118 and EN 1999-1-1) approach the determination of the shear resistance similarly by grouping the sections into non-slender and slender sections. For non-slender sections the shear buckling check of the web parts can be omitted but it must be carried out for the slender sections. Sectional weakening due to holes and HAZ are addressed similarly in both design codes.

Classification for shear buckling
To determine the susceptibility to shear buckling according to EN 1999-1-1 the section must first be classified and defined as either non-slender or slender. This process is also required using BS 8118 where the sections are separated into groups described as either compact or

slender. The main difference between EN 1999-1-1 and BS 8118 in this classification process is the greater depth-to-thickness ratio to define non-slender and slender sections. Ratios for both design codes are given in Table 7.11.

Table 7.11 Slenderness ratios, classification for shear buckling (where $\varepsilon = \sqrt{(250/f_o)}$ and f_o = characteristic value of 0.2% proof strength)

	$\beta =$	**Compact = non-slender**	**Slender**
BS 8118	d/t	$\beta \leq 49\ \varepsilon$	$\beta > 49\ \varepsilon$
EN 1999-1-4	h_w/t_w	$\beta \leq 39\ \varepsilon$	$\beta > 39\ \varepsilon$

Shear area

The definition of the shear area is based on the shape of the cross-section and the shear stress distribution within this shape. A typical shear stress distribution for open cross-sections and solid cross-section is illustrated in Figure 7.17.

As shown in Figure 7.17, the shear stress is not uniformly distributed throughout a cross-section. Thus, when designing for shear, the entire cross-sectional area is not used to calculate the shear resistance. For cross-sections with web parts only these web parts are used, and for solid sections the reduced area of the cross-section is taken. Appropriate equations to determine the relevant shear area are given in both design codes. Equations (7.47)–(7.48), which are used to compute the shear area, are taken from EN 1999-1-1.

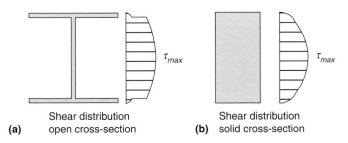

Shear distribution	Shear distribution
(a) open cross-section	**(b)** solid cross-section

Figure 7.17 Typical shear distribution in cross-sections

- Shear area, A_v for sections containing shear webs:

$$A_V = \sum_{i=1}^{n}\left[\left(h_w - \sum d\right)\times(t_W)_i - \left(1 - \rho_{0,haz}\right)\times b_{haz}\times(t_w)_i\right] \qquad (7.47)$$

where

$$h_w = \text{depth of web(s) between the flanges}$$
$$t_w = \text{web thickness}$$
$$d = \text{diameter of holes along shear plane}$$
$$n = \text{number of webs}$$
$$\rho_{o,haz} = \text{reduction factor for HAZ}$$
$$b_{haz} = \text{total depth of HAZ material}$$
$$\text{(see also notes in EN 1999-1-1, Cl.6.2.6)}$$

- Shear area, A_v for solid bar and round tube sections:

$$A_v = \eta_v \times A_e \qquad (7.48)$$

where $\eta_v = 0.8$ for a solid bar and 0.6 for a round tube; and $A_e = $ effective area taking HAZ into consideration.

Non-slender sections
For non-slender section the shear design resistance of a cross-section is given by:

$$V_{Rd} = A_v \times \frac{f_0}{\sqrt{3 \times \gamma_{M1}}} \qquad (7.49)$$

where

$$A_v = \text{shear area}$$
$$f_o = \text{characteristic value of 0.2\% proof strength}$$
$$\gamma_{M1} = \text{partial factor for resistance of cross-section}$$

A design check is not required in order to show adequate resistance to shear buckling.

Slender sections
Flexural members with web parts classified as slender must be checked against the shear buckling of the web parts. When the shear stress reaches a critical value then the web part or web plate buckles. The plate distorts in plane and buckles appear along the tensile direction. However, it is the compression that causes the buckling, as was discussed earlier in this chapter. The principle of shear buckling is shown in Figure 7.18.

The new Eurocode 9 (EN 1999-1-1) also considers the tension field action possible with the use of web stiffeners. The tension field design approach allows the use of the reserve of strength after elastic critical buckling but also requires careful structural design of the member and stiffener plates.

In a state of pure shear stress there will be principal tensile and compressive stresses as shown in Figure 7.18. As discussed, the compressive stresses result in wrinkling or buckling. Due to the buckling the compressive stresses cannot increase further. However, the tensile stresses

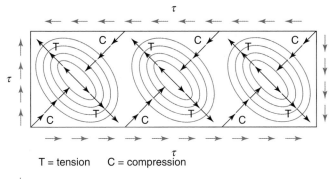

τ

T = tension   C = compression

Figure 7.18 | Shear buckling of web/plate parts

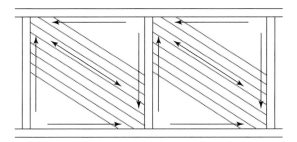

Figure 7.19 | Tension field in post-buckled shear web

will increase with added shear force. These increased tensile stresses act diagonally and form the tension fields. This behaviour forms a truss system of forces, where the flanges resist the pulling-in/bending effect of the tension field and transverse stiffeners act as struts supporting the flanges. The ultimate shear capacity is eventually reached due to the tensile yielding forming plastic hinges in the flanges (see Figure 7.19)

For a web panel the critical (elastic) shear buckling stress is given by:

$$\tau_{cr} = k_{\tau} \times \frac{\pi^2 \times E}{12(1 - v^2)} \times \frac{t_W^2}{b_W^2} \tag{7.50}$$

where

E = modulus of elasticity
v = Poission's ratio
t_w = web thickness
b_w = web depth
k_{τ} = buckling coefficient for shear buckling

As the buckling depends on the length of the element and support conditions, the buckling coefficient for shear buckling k_{τ} is utilised to account for some these factors. Again, both design codes BS 8118 and EN 1999-1-1 provide guidance and equations to determine the shear buckling coefficient. Equations (7.51) and (7.52), taken from Eurocode 9 (EN 1999-1-1), give the shear buckling factor k_τ:

$$\text{for ratio } a/b \geq 1.00 \quad k_\tau = 5.34 + [4.00 \times (b/a)^2] \qquad (7.51)$$

$$\text{for ratio } a/b < 1.00 \quad k_\tau = 4.00 + [5.34 \times (b/a)^2] \qquad (7.52)$$

where a = longitudinal distance between plate restraints and b = transverse depth between plate restraints.

In addition to the shear buckling coefficient, EN 1999-1-1 applies a reduction factor for shear buckling v_1. This procedure can also be found in the method to design plate girders, where the factor for shear buckling, ρ_v is dependent on the support conditions and slenderness parameter for shear buckling.

The reduction factor for shear bucking v_1 is calculated by:

$$v_1 = \frac{17 \times t \times \varepsilon \times \sqrt{k_\tau}}{b} \qquad (7.53)$$

but not more than:

$$v_1 = k_\tau \times \frac{430 \times t^2 \times \varepsilon^2}{b^2} \qquad (7.54)$$

where

t = thickness of cross-section part
b = transverse depth between plate restraints
ε = coefficient = $\sqrt{(250/f_o)}$
k_τ = buckling coefficient for shear buckling

For slender plates with a ratio $\beta > 39 \times \varepsilon$, the values for both the design shear resistance for the yielding and buckling must be checked. For shear yielding the procedure has been laid out earlier in this chapter and is the same as for non-slender sections.

$$V_{Rd} = A_v \times \frac{f_0}{\sqrt{3 \times \gamma_{M1}}} \qquad (7.55)$$

The design value for resisting shear buckling is given by:

$$V_{Rd} = v_1 \times b \times t \times \frac{f_0}{\sqrt{3 \times \gamma_{M1}}} \qquad (7.56)$$

where

v_1 = reduction factor for shear buckling
b = transverse depth between plate restraints
t = thickness of cross-section part
f_o = characteristic value of 0.2% proof strength
γ_{M1} = partial factor for resistance of cross-section

It should be noted that holes and cut-out in the shear area, i.e. the web section, should be small and limited to maximum diameters of $d/h_w \leq 0.05$.

7.7.1.4 Bending and shear

In flexural members bending moment and shear act at the same time, thus the shear stresses that are present will have an effect on the moment capacity of a flexural member. Whilst the resistance to bending moments is often the foremost criteria for the design of flexural members, the presence of shear might reduce the moment capacity. However, if the shear force V_{Ed} is less than half the shear resistance V_{Rd} its effect on the moment capacity may be neglected. This does not apply where shear buckling reduces the resistance of the section.

Figure 7.20 illustrates the interaction of shear force and bending moment resistance. Further information and guidance can be obtained from BS EN 1999-1-1 Clause 6.7.7 Interaction.

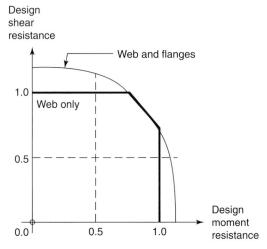

Figure 7.20 | Interaction of shear resistance and moment resistance

For cases where it is necessary to reduce the member's moment resistance the design value of the material strength is reduced by:

$$f_{o,V} = f_o \times \left(1 - \left(\frac{2 \times V_{Ed}}{V_{Rd}} - 1\right)^2\right) \qquad (7.57)$$

where

$f_{o,V}$ = reduced design value of strength
f_o = characteristic value of 0.2% proof strength
V_{Ed} = design shear force
V_{Rd} = design shear resistance

Equation 7.57 is applicable in general, nevertheless, some special rules laid out in EN 1999-1-1 might be beneficial when designing cross-sections in which the conditions given below are applicable:

- For an equal flanged I-section classified as Class 1 or 2 in bending, the design value of the reduced moment resistance due to the presence of shear is given by:

$$M_{V,Rd} = t_f \times b_f \times (h - t_f) \times \frac{f_o}{\gamma_{M1}} + \frac{t_w \times h_w^2}{4} \times \frac{f_{o,V}}{\gamma_{M1}} \qquad (7.58)$$

- For an equal flanged I-section classified as Class 3 in bending, the design value of the reduced moment resistance due to the presence of shear is given by:

$$M_{V,Rd} = t_f \times b_f \times (h - t_f) \times \frac{f_o}{\gamma_{M1}} + \frac{t_w \times h_w^2}{6} \times \frac{f_{o,V}}{\gamma_{M1}} \qquad (7.59)$$

where

$M_{v,Rd}$ = reduced moment resistance
t_f = flange thickness
t_w = web thickness
b_f = flange width
h = total depth of section
h_w = depth of web between inside flanges
f_o = characteristic value of 0.2% proof strength
$f_{o,V}$ = reduced design value of strength
γ_{M1} = partial factor for resistance of cross-section

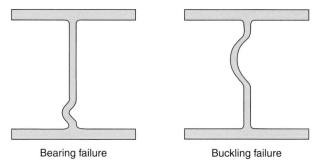

Bearing failure   Buckling failure

Figure 7.21 | Web bearing and buckling failure

7.7.1.5 *Web bearing*

Web bearing failure can occur at locations where concentrated loads are applied to the top or bottom flange i.e. at supported point loads or the supports of a beam member. The web or webs must be checked to ensure that the web part of the section is sufficient to carry and disperse these loads. The two possible failure modes are (see Figure 7.21):

- web bearing failure
- web buckling failure

The I-section used for the presentation in Figure 7.21 is typically only as the two failure modes can equally occur in multi-web sections.

Bearing failure
Web bearing failure occurs due to local crushing of the web at the location of the concentrated load. Material yielding at the point of the concentrated load causes failure of the web closest to the loaded length of the web. Crushing of the web will also allow deformation of the adjacent flange.

Buckling failure
Web buckling failure can be localised by buckling of the web adjacent to the concentrated load or support. But it can also occur over a longer part of the web if multiple point loads or a distributed load are applied to the flange.

It should be noted that web bearing and web buckling design checks are usually not required if the load is directly applied to the web part i.e. by end plate, cleats, fin plate or similar connections.

The resistance to web bearing/web buckling depends on a few parameters and factors such as:

- effective loaded length
- support conditions and restraints due to stiffeners
- web slenderness

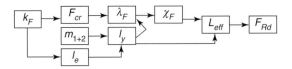

k_F = buckling factor for transverse loads

F_{cr} = elastic critical buckling load

λ_F = slenderness parameter

m_1 = parameter for effective loaded length

m_2 = parameter for effective loaded length

χ_F = reduction factor

l_e = parameter for effective loaded length

l_y = effective loaded length

L_{eff} = effective length

F_{Rd} = design resistance to transverse force

Figure 7.22 | Web bearing – Design procedure to EN 1999-1-1

The method used below to determine the resistance to web bearing and/or web buckling is based on the guidance given in EN 1999-1-1 Clause 6.7.5 *Resistance to transverse loads*, where a number of factors must be calculated first. The Eurocode 9 procedure takes transverse and longitudinal web stiffeners into consideration as well as unstiffened webs. The design procedure for web bearing is termed local buckling in the new Eurocode 9. This method is shown graphically in Figure 7.22.

Buckling factor for transverse loads, k_F
For webs without longitudinal stiffeners, the buckling factor k_F depends on the applied loads and supports. The appropriate equations are given below with additional assistance available in BS EN 1999-1-1 Clause 6.7.5.3.

The design code (EC9) differentiates between three conditions when calculating the buckling factor k_F.

(i)
$$k_F = 6 + 2 \times (b_W / a)^2 \qquad (7.60)$$

for cases:

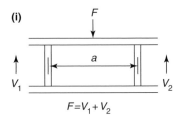

(ii)
$$k_F = 3.5 + 2 \times (b_w / a)^2 \qquad (7.61)$$

for cases:

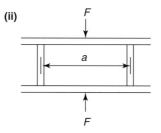

(ii)

(iii)

$$k_F = 2 + 6 \times \left(\frac{S_S + C}{b_W} \right) \le 6 \qquad (7.62)$$

for cases:

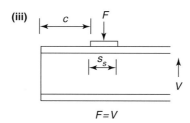

(iii)

where

b_w = distance between fillets (see Figure 7.23 for terminology)
a = longitudinal distance between web restraints (see Figure 7.23 for terminology)
c = beam end distance to 'length of stiff bearing'
s_s = length of stiff bearing

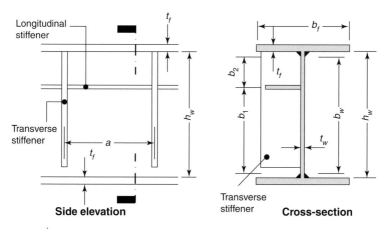

Figure 7.23 | Terminology for stiffened web cross-sections

For webs with longitudinal stiffeners the buckling factor k_F is calculated from:

$$k_F = 6 + 2 \times \left(\frac{h_w}{a}\right)^2 + \left(\frac{5.44 \times b_1}{a} - 0.21\right) \times \sqrt{\gamma_s} \qquad (7.63)$$

where

h_w = depth of web between inside flanges*
a = longitudinal distance between web restraints*
b_1 = vertical distance between web restraint(s)*
γ_s = relative second moment of area

*see Figure 7.23 for terminology

Elastic critical buckling load, F_{cr}
The elastic critical buckling load F_{cr} is based on the discussion earlier with regard to compression members or compression parts and can be found by calculating:

$$F_{cr} = 0.9 \times k_F \times \frac{E \times t_w^3}{h_w} \qquad (7.64)$$

where

E = modulus of elasticity
t_w = web thickness
h_w = depth of web between inside flanges
k_F = buckling factor for transverse loads

Parameters m_1 and m_2 for effective loaded length
Two dimensionless parameters are used to obtain the loaded length l_y. These parameters are denoted m_1 and m_2. Equations (7.65) and (7.66) are extracted from Clause 6.7.5.5 of EN 1999-1-1:

$$m_1 = \frac{f_{of} \times b_f}{f_{ow} \times t_w} \qquad (7.65)$$

where f_{of} = characteristic value of strength for flange and f_{ow} = characteristic value of strength for web:

$$m_2 = 0.02 \times \left(\frac{h_w}{t_f}\right)^2 \quad \text{if } \overline{\lambda}_F > 0.5, \text{ otherwise } m_2 = 0 \qquad (7.66)$$

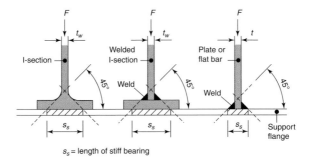

s_s = length of stiff bearing

Figure 7.24 | Typical length of stiff bearing

It should be noted that for box girders, b_f in Equation (7.65) is limited to $15 \times t_f$ on each side of the web.

Parameter for effective loaded length, l_e
The parameter for the effective loaded length l_e is given by:

$$l_e = \frac{k_F \times E \times t_w^2}{2 \times f_{ow} \times h_w} \le s_s + c \tag{7.67}$$

where c = distance to unstiffened end (see Figure 7.24 for clarification) and s_s = length of stiff bearing.

Figure 7.24 indicates the typical length of stiff bearings. It is assumed that the angle of dispersion is accepted at 45° (see BS EN 1999-1-1, Clause 6.7.5.3, Figure 6.31).

If a number of point loads are narrowly spaced, the design resistance should be checked for each individual load and the total load with the length of stiff bearing s_s taken as the centre distance measured between the outmost loads.

Effective loaded length, l_y
The effective loaded length l_y should be calculated using the parameters m_1 and m_2 according to the cases shown below:

(i) $$l_y = s_s + 2 \times t_f \times \left(1 + \sqrt{m_1 + m_2}\right), \text{ but } l_y \le a \tag{7.68}$$

for cases:

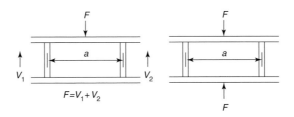

For the case illustrated below, l_y should be obtained as the smaller of the results for the following equations:

(ii)

$$l_y = l_e + t_f \times \sqrt{\frac{m_1}{2} + \left(\frac{l_e}{t_f}\right)^2} + m_2 \qquad (7.69)$$

$$l_y = l_e + t_f \times \sqrt{m_1 + m_2} \qquad (7.70)$$

for cases:

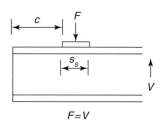

$$F = V$$

The effective loaded length l_y is subsequently utilised to determine the slenderness parameter λ_F and furthermore the effective length L_{eff}.

Slenderness parameter, λ_F
The slenderness parameter λ_F for local buckling due to transverse load can be obtained from:

$$\lambda_F = \sqrt{\frac{F_R}{F_{cr}}} = \sqrt{\frac{l_y \times t_w \times f_{ow}}{F_{cr}}} \qquad (7.71)$$

where

l_y = effective loaded length
t_w = web thickness
f_{ow} = characteristic value of strength for web
F_{cr} = elastic critical buckling load

Reduction factor, χ_F
The reduction factor χ_F for resistance is obtained from:

$$\chi_F = \frac{0.5}{\lambda_F} \quad \text{but should not exceed 1.0} \qquad (7.72)$$

Effective length, L_{eff}
The effective length L_{eff} is the product of the effective loaded length l_y and the reduction factor χ_F. It is given by:

$$L_{eff} = \chi_F \times l_y \qquad (7.73)$$

Design resistance to transverse force, F_{Rd}

For unstiffened and stiffened webs the design resistance F_{Rd} to local buckling i.e. web bearing at transverse loads is given by:

$$F_{Rd} = \frac{L_{eff} \times t_w \times f_{ow}}{\gamma_{M1}} \qquad (7.74)$$

where

L_{eff} = effective length
t_w = web thickness
f_{ow} = characteristic value of strength for web
γ_{M1} = partial factor for resistance of cross-section

It should be noted that box girders with inclined webs, webs and associated flange should be checked. The load effects are the parts of the external load in the plane of webs and flange, respectively. The effect on the transverse force on the moment resistance of the affected member should also be checked. Figure 7.25 illustrates the compressive forces in plane of web and flange parts.

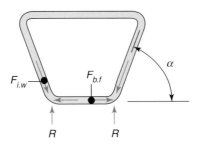

Figure 7.25 | Members with inclined webs

The forces in webs and flanges of box girders with in inclined webs can be calculated by applying the trigonometrical functions as shown below.

- For inclined webs:

$$F_{i.w} = R/\sin(\alpha) \qquad (7.75)$$

- For bottom flanges:

$$F_{b.f} = R/\tan(\alpha) \qquad (7.76)$$

where

$F_{i.w}$ = force acting in inclined webs
$F_{b.f}$ = force acting in bottom flange
R = support reaction
α = angle of inclination
(see Figure 7.25 for notation)

7.7.1.6 Lateral-torsional buckling

When a beam is loaded in flexure, the acting bending moment will cause compression in the upper part and tension in the lower part of the beam. As the load acting on the beam is increased close to or beyond the critical elastic load, the unrestrained compression part that is free to move sideways will buckle. The tension part that is being pulled along will try to resist this sideways movement. This interaction between compression and tension parts resulting in beam buckling and twisting, will leave the web of an I-beam non-vertical. This distortion is known as lateral-torsional buckling (LTB) in structural design. With little or no lateral restraints along a slender beam LTB may be the dominant design criteria as the distorted beam will only have reduced capacities compared to a straight beam. Both BS 8118 and Eurocode 9 give a method to compute the reduced capacity of flexural members susceptible to LTB buckling.

Figure 7.26 shows the typical buckling i.e. curvature of the top flange and no longer vertical web of the cross-section. Flexural members i.e. beams and beam-columns with little torsional and lateral stiffness such as slender I-beams and channels, and also rectangular solid or hollow sections with a deep and narrow cross-section, are the most vulnerable to failure due to LTB. However, there are also exceptions where LTB does not need to be checked. No LTB checks are required for the following conditions/criteria:

- the member is fully restraint against lateral movement throughout its length
- bending takes place about the minor axis
- the relative slenderness $\bar{\lambda}_{LT}$ between points of effective restraint is less than 0.4
- for rectangular hollow sections where $h/b < 2$
- for circular hollow sections

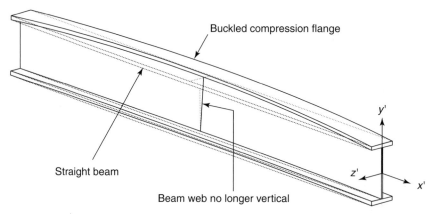

Buckled compression flange

Straight beam

Beam web no longer vertical

y'

z'

x'

Figure 7.26 | Lateral-torsional buckling of an I-beam section

Elastic critical moment for lateral-torsional buckling, M_{cr}
When designing to prevent LTB buckling of flexural members, it is first necessary to determine the elastic critical moment M_{cr}. This is often a complex process as the elastic critical moment depends on numerous conditions, such as:

- material properties
- geometrical properties of the cross-section
- torsion and warping rigidity of the cross-section
- torsional and warping restraints of the cross-section at supports
- position of the transverse loading in relation to the shear centre
- moment gradient

A beam formula to calculate the elastic critical moment and guidance in applying the equations and coefficients is given in Annex I.1 of EN 1999-1-1.

The elastic critical moment for LTB of a beam of uniform symmetrical cross-section with equal flanges, under standard conditions of restraint at each end and subject to uniform moment in plane going through the shear centre, is given by (see EN 1999-1-1, Annex I.1.1):

$$M_{cr} = \frac{\pi \times \sqrt{E \times I_z \times G \times I_t}}{L} \times \sqrt{1 + \frac{\pi^2 \times E \times I_w}{L^2 \times G \times I_t}} \qquad (7.77)$$

The elastic critical moment for LTB of a beam of uniform cross-section which is symmetrical about the minor axis, for bending about the major axis is obtained from (see EN 1999-1-1, Annex I.1.2):

$$M_{cr} = \mu_{cr} \times \frac{\pi \times \sqrt{E \times I_z \times G \times I_t}}{L} \qquad (7.78)$$

where

I_t = torsion constant of cross-section
I_z = second moment of area about minor axis
I_w = warping constant
v = Poisson ratio
L = length between points of lateral restraint

Relative slenderness parameter, $\bar{\lambda}_{LT}$
The relative slenderness parameter $\bar{\lambda}_{LT}$ is given by:

$$\bar{\lambda}_{LT} = \sqrt{\frac{\alpha \times W_{el,y} \times f_o}{M_{cr}}} \qquad (7.79)$$

where

α = shape factor, but $\alpha \leq W_{pl,y}/W_{el,y}$
$W_{el,y}$ = elastic section modulus of gross cross-section
f_o = characteristic value of 0.2% proof strength
M_{cr} = elastic critical moment for LTB

Similarly, the relative slenderness parameter $\bar{\lambda}_{LT}$ can be calculated without determining the elastic critical moment for LTB, M_{cr}, by using the method given in Annex I.2 of EN 1999-1-1. For I-sections and channels complying with the criteria set out in Table 7.12, the relative slenderness parameter can be found by:

$$\bar{\lambda}_{LT} = \lambda_{LT} \times \frac{1}{\pi} \times \sqrt{\frac{\alpha \times f_o}{E}} \qquad (7.80)$$

where

α = shape factor, but $\alpha \leq W_{pl,y}/W_{el,y}$
f_o = characteristic value of 0.2% proof strength
E = modulus of elasticity
λ_{LT} = slenderness parameter for LTB (see below)

$$\lambda_{LT} = \frac{X \times L_{cr,z}/i_z}{\left[1 + Y \times \left(\dfrac{L_{cr,z}/i_z}{h/t_2}\right)^2\right]^{1/4}} \qquad (7.81)$$

where

X = LTB coefficient
Y = LTB coefficient
$L_{cr,z}$ = buckling length for LTB
i_z = radius of gyration of cross-section
h = overall section depth
t_2 = flange thickness

It should be noted that conservative values of $X = 1.0$ and $Y = 0.05$ could be taken for all cases of X and Y.

Table 7.12 | Imperfection factors to EN 1999–1–1

Cross-section class	α_{LT}	$\bar{\lambda}_{0,LT}$
1 + 2	0.10	0.60
3 + 4	0.20	0.40

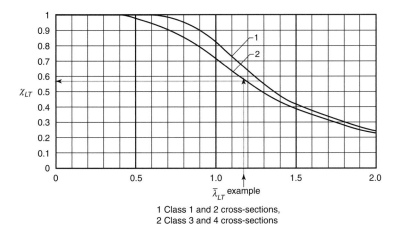

1 Class 1 and 2 cross-sections,
2 Class 3 and 4 cross-sections

Figure 7.27 | Reduction factor χ_{LT} (Extracted from EN 1999-1-1, Figure 6.13)

Reduction factor for LTB, χ_{LT}

The reduction factor χ_{LT} for torsional-flexural buckling can be obtained directly from Figure 7.27 extracted from Eurocode 9 where the arrows indicate an arbitrary example.

However, the value of the reduction factor χ_{LT} can also be determined numerically by Equation (7.82), which has been extracted from EN 1999-1-1 Clause 6.3.2.2:

$$\chi_{LT} = \frac{1}{\phi_{LT} + \sqrt{\phi_{LT}^2 - \bar{\lambda}_{LT}^2}} \, but \, \chi_{LT} \leq 1 \qquad (7.82)$$

where:

$$\phi_{LT} = 0.5 \times \left[1 + \alpha_{LT} \times \left(\bar{\lambda}_{LT} - \bar{\lambda}_{0,LT} \right) + \bar{\lambda}_{LT}^2 \right] \qquad (7.83)$$

α_{LT} = imperfection factor (see Table 7.12)
$\bar{\lambda}_{LT}$ = relative slenderness
$\bar{\lambda}_{0,LT}$ = limit of horizontal plateau (see Table 7.12)

The values of the imperfection factor α_{LT} and $\bar{\lambda}_{0,LT}$ are taken from Table 7.12.

Design buckling resistance, $M_{b,Rd}$

The design buckling resistance moment $M_{b,Rd}$ of laterally unrestrained members should be calculated from:

$$M_{b,Rd} = \chi_{LT} \times \alpha \times W_{el,y} \times f_o / \gamma_{M1} \qquad (7.84)$$

where

χ_{LT} = reduction factor for LTB
α = shape factor, but $\alpha \leq W_{pl,y} / W_{el,y}$

$W_{el,y}$ = elastic section modulus of gross cross-section
f_o = characteristic value of 0.2% proof strength
γ_{M1} = partial factor for resistance of cross-section

Figure 7.28 illustrates lateral-torsional failure of an I-beam section overloaded during buckling analysis under standard conditions of restraint at each end and subject to uniform moment in plane going through the shear centre.

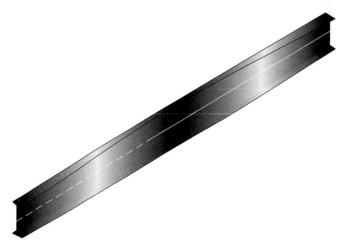

Figure 7.28 | LTB failure of I-section

7.7.2 Bending and axial force members

It is not unusual for flexural members to also be loaded axially. A typical example is beam-columns which provide lateral rigidity and also carry gravity loads. The interaction of bending moment, shear force and axial force must be considered and affected members should be checked for the combined stresses. The effect of the combination of the bending and axial stresses are best demonstrated by the simplified graphic shown in Figure 7.29.

A simple, conservative approach to checking for combined resistance to bending and axial force would be to add the utilisation ratios for the moment and axial force. This method, based on Figure 7.29, would check the increased stress in one of the flanges by adding the stress introduced by the axial load. This technique is used in BS 8118:

$$\frac{M_{Ed}}{M_{Rd}} + \frac{N_{Ed}}{N_{Rd}} \leq 1.0 \qquad (7.85)$$

where

M_{Ed} = design moment
M_{Rd} = moment resistance

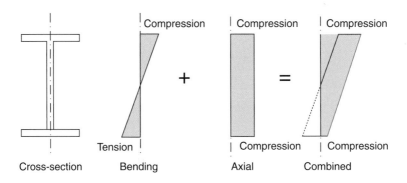

Figure 7.29 | Combined moment and axial force

N_{Ed} = design axial force

N_{Rd} = axial force resistance

However, the method described in EN 1999-1-1 is more clearly defined and additional factors are added to determine the adequacy of a member subjected to bending and axial load. The factors used to check structural sufficiency of a member are added to consider the effect of welding and cross-sectional shape.

For doubly symmetrical cross-sections the following criteria must be satisfied:

$$\left(\frac{N_{Ed}}{\omega_0 \times N_{Rd}}\right)^{\eta_0} + \left(\frac{M_{y,ED}}{\omega_0 \times M_{y,Rd}}\right)^{\gamma_0} + \left(\frac{M_{z,Ed}}{\omega_0 \times M_{z,Rd}}\right)^{\xi_0} \le 1.00 \qquad T(7.86)$$

where

$\eta_0 = 1.0$ or may alternatively taken as $\alpha_z^2 \times \alpha_y^2$ but $1 \le \eta_0 \le 2$
$\gamma_0 = 1.0$ or may alternatively taken as α_z^2 but $1 \le \gamma_0 \le 1.56$
$\xi_0 = 1.0$ or may alternatively taken as α_z^2 but $1 \le \xi_0 \le 1.56$

For solid and hollow cross-sections the criteria given in Equation (7.87) must be satisfied:

$$\left(\frac{N_{Ed}}{\omega_0 \times N_{Rd}}\right)^{\psi} + \left[\left(\frac{M_{y.Ed}}{\omega_0 \times M_{y,Rd}}\right)^{1.7} + \left(\frac{M_{z,Ed}}{\omega_0 \times M_{z,Rd}}\right)^{1.7}\right]^{0.6} \le 1.00 \qquad (7.87)$$

where

$\psi = 1.3$ for hollow sections or optional taken as $\alpha_z \times \alpha_y$ but $1 \le \psi \le 1.3$

$\psi = 2.0$ for solid sections or optional taken as
$\alpha_z \times \alpha_y$ but $1 \le \psi \le 2.0$
α = shape factor of cross-section
ω = HAZ factor (see Section 7.7.1.2)
N_{Ed} = design axial force
N_{Rd} = axial force resistance
$M_{y,Ed}$ = design moment about y-axis
$M_{z,Ed}$ = design moment about z-axis
$M_{y,Rd}$ = moment resistance about y-axis
$M_{z,Rd}$ = moment resistance about z-axis

Example 7.5: Simple beam design
Check the bending and shear resistance of the I-section used in Examples 7.1 and 7.4 earlier in this chapter. There is no welding or holes along the section.

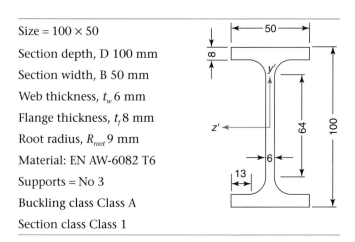

Size = 100×50
Section depth, D 100 mm
Section width, B 50 mm
Web thickness, t_w 6 mm
Flange thickness, t_f 8 mm
Root radius, R_{root} 9 mm
Material: EN AW-6082 T6
Supports = No 3
Buckling class Class A
Section class Class 1

Criteria: UDL loading = 5.000 kN/m; effective span = 2.000 m; and section class = 1 (web and flange)

Example 7.1
Material data   (EN 1999-1-1, Table 3.2b)

For EN AW-6082 T6 EP/O
$f_o = 260$ N/mm²
$f_u = 310$ N/mm²

Partial safety factors   (EN 1999-1-1, 6.1.3)

$\gamma_{M1} = 1.10$
$\gamma_{M2} = 1.25$

(1) Analysis of beam:

Design bending moment, $M_{Ed} = 10 \times 2/8 = $ **2.500** kNm
Design shear, $V_{Ed} = 10/2 = $ **5.000** kN

(2) Bending moment resistance   (EN 1999-1-1, 6.2.5)

$W_{net} = 42.099$ cm³
$W_{el} = 42.099$ cm³
$W_{pl} = 50.169$ cm³
$\alpha = 50.169/42.099 = 1.192$ (EN 1999-1-1, Table 6.4)
$M_{u,Rd} = 42099 \times 310/1.25 = 10\,440\,552$ Nmm > 2.500 kNm
$M_{c,Rd} = 1.192 \times 42099 \times 260/1.10 = 11\,861\,202$ Nmm > 2.5 kNm

Thus the resistance to pure bending would be 10.441 kNm.

(3) Shear resistance   (EN 1999-1-1, Table 3.2b)

$\varepsilon = (250 / 260)^{0.5} = 0.981$
$39 \times \varepsilon = 39 \times 0.981 = 38.3$
$h_w = 100 - 16 = 84$ mm
$t_w = 6$ mm
$h_w/t_w = 84 / 6 = 14.0 < 38.3$ (EN 1999-1-1, 6.2.6(2))
$A_{net} = 84 \times 6 = 504$ mm²
$V_{Rd} = 504 \times 260/(\sqrt{3} \times 1.1) = $ **68 778** N > 5.000 kN

(4) Bending and shear   (EN 1999-1-1, 6.2.8)

$V_{Rd} \geq 2 \times V_{Ed} = 68.778 \geq 10.000$ | true

Low shear, hence the effect on the moment resistance can be
neglected   (EN 1999-1-1, 6.2.8(2))

(5) Buckling resistance   (EN 1999-1-1, 6.3.2.1)
Member laterally unrestrained

$\alpha = 1.192$ (EN 1999-1-1, Table 6.4)
$L = 2000$ mm
$I_z = 17.013$ cm⁴
$I_t = 3.234$ cm⁴
$I_w = 0.360 \times 10^9$ mm⁶
$G = 27\,000$ N/mm²   (EN 1999-1-1, Annex I.1)
$E = 70\,000$ N/mm²
$M_{cr} = 6.413$ kNm   (EN 1999-1-1, Annex I.1)
$\overline{\lambda} = [(1.192 \times 42099 \times 260)/6\,413\,130]^{0.5} = 1.426$
$\phi_{LT} = 0.5 \times [1 + 0.1 \times (1.426 - 0.6) + 1.426^2] = 1.558$
$\chi_{LT} = 1/(1.558 + (1.558^2 - 1.426^2)^{0.5} = 0.458 \leq 1.0$ | true
$M_{b,Rd} = 0.458 \times 1.192 \times 42099 \times 260/1.1 = 5432430.466$ Nmm
$M_{b,Rd} = 5.432$ kNm > 2.500 kNm | true

(6) Solution: The 100×50 I-section beam with a span of 2000 mm
loaded uniformly with 5.00 kN/m design load by inspection is
satisfactory.

8

CONNECTIONS

Introduction

Connections are an important part of every structure. The design and specification of the connections, also known as joints, of a structure is an inherent part of the structural design process. Nevertheless, designing the connections is often passed by project engineers to specialist contractors or other engineers. This is due to complexity and comprehension of involved design consideration and experience necessary.

Accurate information and knowledge of how the loadings are transferred between structural members and correct determination of the load paths form the basic principles of connection design. However, exact interpretation of the structural analysis results and right specification within the structural analysis is equally important to obtain connections that are structurally safe but also economic to fabricate.

Guidance for designing joints for aluminium structures is given in Eurocode 9 (EN 1999–1) Clause 8 *Design of joints*. This chapter does not intend to provide exhaustive guidance for designing connections nor will it provide a detailed exploration of the structural mechanics acting on joints. However, it should provide some basic insights into the design procedure and structural requirements when designing connections.

For structural design connections are grouped into the following connection types:

- mechanical joints
- welded joints
- bonded joints

Whilst, for simplicity of design it is desirable to maintain the same connection type throughout the entire joint, a combination of mechanical and welded connections is often used.

Figure 8.1 illustrates the most common types of connections. At a first glance the fully welded connection seems to be the simplest and

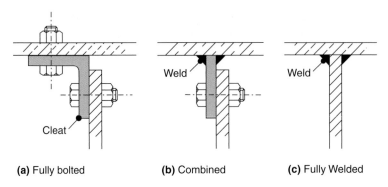

(a) Fully bolted   **(b)** Combined   **(c)** Fully Welded

Figure 8.1 | Typical joint types

most cost-effective to fabricate. However, erection and transportation requirements, and also the effects of the HAZ may well prohibit the use of welding. In contrast, due to the lightweight property of aluminium small structures could be fully welded but limited by transportation problems due to their overall size rather than by lifting restrictions.

A first overview of jointing for aluminium members has been given in Section 2.6.4 of this book, where the methods of joining were described. As with the virtually unlimited number of cross-section shapes, which often provide added functionality for fixing and joining, the possibilities of joining extruded aluminium members is infinite. The most common types of joint that are used in the construction industry are highlighted in this chapter.

8.1 Mechanical connections

Mechanical connections formed by bolting, screwing, riveting and pinning are frequently used as connection methods when joining aluminium members. Despite their simplicity, the mechanical connections require considerations beyond the structural integrity of the joint. Material selection and detailing for all of the parts used to form the connection is important for preventing corrosion, as will be discussed in Chapter 9.

8.1.1 Design principle

The principle that applies to the design of structural members, as discussed earlier in this book, also applies to the design of connections. It is necessary to verify that the design resistance is greater than the design force for each part of the connection as illustrated in the following basic equation:

$$\frac{F_{Ed}}{F_{Rd}} \leq 1.0 \tag{8.1}$$

where F_{Ed} = design force and F_{Rd} = design resistance. This design requirement applies to all types of bolted joints (i.e. mechanical, welded or adhesively bonded connections). During the structural design of joints it is important to determine the accurate load path and possible modes of failures. each possible failure along the load path, the condition given by Equation (8.1) must be checked and confirmed for each possible failure along the load path.

8.1.2 Load path

The load path describes how the load is transferred through the joint, i.e. from member to member via connection elements such as bolts, cleats and plates. Once identified, the load path lists all involved parts forming the load path. All these parts listed by the load path must be checked for structural strength and integrity and also for compliance with the applicable design codes.

Figure 8.2 shows a load path for a typical bolted joint where an angle cleat is utilised to join the beam to the column. The load is transferred from the beam① via bolts②, angle cleat③ and again bolts④ to the column⑤. The failures that may occur along this load path can then be identified.

For more complex joints it may be beneficial to use load path diagrams or load path tables which list the parts that are involved and the possible failure modes. An example of a load path table is shown in Table 8.1 which indicates the complexity involved in designing joints. Thus, detailed discussion and step-by-step guidance of design checks and procedures would extend beyond the scope of this chapter and

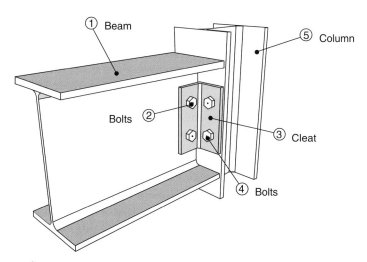

Figure 8.2 | Example of load path for bolted joint

Table 8.1 | Load path table (example)

Path No.	Member/part	Failure mode	Notes
1	Beam	Shear	
		Bending	Reduced
		Axial	due holes
		Combination	
		Bearing in holes	
		Integrity	Tying force
2	Bolts	Shear	
		Bearing	
3	Cleat	Bearing in holes	
		Plain shear	
		Block tearing	
		Cleat deformation	Tying force
4	Bolts	Shear	
		Bearing	
		Tension	
5	Column	Bearing in holes	
		Local shear	
		Axial	
		Integrity	Tying force

book. Directions for designing joints are given by the relevant design codes, for example Eurocode 9 (EN 1999–1) Clause 8 *Design of joints*.

8.1.3 Fastener material

Possible fastener materials for use in aluminium and aluminium alloy structures include:

- aluminium/aluminium alloys
- steel (mild steel)
- stainless steel

Despite the choices available for fastener materials, it is recommended that fasteners manufactured from aluminium alloy material should

Table 8.2 | Coefficients of thermal expansion

Aluminium	Mild steel	Stainless steel
$23.5 \times 10^{-6}/°C$	$12 \times 10^{-6}/°C^*$	$16 \times 10^{-6}/°C^*$

*depends on composition of material

be used. This is in order to avoid bi-metallic corrosion i.e. being in contact with more noble materials and also thermal expansion problems. Loss of tightness can occur when mild steel or stainless steel fasteners are utilised for joints between aluminium elements. The difference in thermal expansion of the two materials will lead to loose connections after many cycles of temperature change (see Table 8.2).

Although aluminium fasteners have a slight shortcoming in terms of strength when compared to mild steel and stainless steel fasteners, they offer a range of compensating advantages. These include:

• resistance to corrosion
• cheaper than stainless steel fasteners
• compatibility to joint elements
• variety of available fasteners
• good conductivity (if desired)
• strength
• colourability (available in many colours and finishes)

Aluminium made bolts and nuts are available in all common metric and imperial sizes. A great variety of rivets, screws and specialised non-standard fasteners are readily obtainable from many manufacturers. Aluminium fasteners can be obtained in a number of alloys and can also be manufactured for the desired purpose or special structural requirements (see Table 8.3).

Table 8.3 | Aluminium alloys for bolts/nuts

Alloy designation	Tensile strength	Shear strength
	(N/mm²)	(N/mm²)
2024-T4	460	285
5056-H18	435	235
6061-T6	310	190
6063-T6	245	150
6082-T6	340	210
6262-T9	400	240
7075-T6	570	350

8.1.4 Bolt capacities

The advantages of aluminium bolts were shown in Section 8.1.3. However, in spite of these advantages it must be admitted that bolts manufactured from aluminium alloys have less resistance to tension, shear and bearing than bolts made from steel and/or stainless steel. The guidance and methodology provided by Eurocode 9 (EN 1999–1-1) should be employed to determine the capacities of the bolt. To verify the adequacy of proposed bolts and other fasteners it is necessary to check whether or not the capacities of the fastener are greater than the applied design actions. The capacities listed below must be calculated and checked against the appropriate design actions:

- shear capacity:

$$F_{v,Rd} = \alpha_v \times f_{ub} \times A / \gamma_{M2} \tag{8.2}$$

- tension capacity:

$$F_{t,Rd} = k_2 \times f_{ub} \times A_s / \gamma_{M2} \tag{8.3}$$

- combined shear and tension capacity:

$$\frac{F_{v,Ed}}{F_{v,Rd}} + \frac{F_{t,Ed}}{1.4 \times F_{t,Rd}} \leq 1 \tag{8.4}$$

- bearing capacity:

$$F_{b,Rd} = k_1 \times \alpha_b \times f_u \times d \times t / \gamma_{M2} \tag{8.5}$$

The factors and criteria can be obtained from EN 1999–1–4 Table 8.5.

Table 8.4 compares the capacities of bolts to Eurocode 9 (EN 1999–1–1) for bolts manufactured from steel, stainless steel and aluminium alloy material.

8.2 Welded connections

Aluminium structural elements are often joined together by welding. Figure 8.1 shows the simplicity of a welded connection. Their advantages are summarised below:

- simplicity of connections and design
- less material required compared to bolted connections
- saving of labour i.e. fabrication time
- less space required than for bolted connections
- no section reduction due to holes
- avoidance of crevices and associated corrosion

Table 8.4 | Bolt capacities to EC9

Diameter	Material	Alloy/grade	A$_s$ (mm²)	Shear (kN)	Tension (kN)
M12	Steel	4.6	84.3	16.186	24.278
		8.8	84.3	32.371	48.557
		10.9	84.3	33.720	60.696
	Stainless steel	50	84.3	16.860	30.348
		70	84.3	23.604	42.487
		80	84.3	26.976	48.557
	Aluminium alloy	5019	84.3	10.453	10.453
		6082	84.3	10.453	10.453
M16	Steel	4.6	157	30.144	45.216
		8.8	157	60.288	90.432
		10.9	157	62.800	113.040
	Stainless steel	50	157	31.400	56.520
		70	157	43.960	79.128
		80	157	50.240	90.432
	Aluminium alloy	5019	157	17.584	17.584
		6082	157	19.468	19.468
M20	Steel	4.6	245	47.040	70.560
		8.8	245	94.080	141.120
		10.9	245	98.000	176.400
	Stainless steel	50	245	49.000	88.200
		70	245	68.600	123.480
		80	245	78.400	141.120
	Aluminium alloy	5019	245	27.440	27.440
		6082	245	30.380	30.380
M24	Steel	4.6	353	67.776	101.664
		8.8	353	135.552	203.328
		10.9	353	141.200	254.160
	Stainless steel	50	353	70.600	127.080
		70	353	98.840	177.912
		80	353	112.960	203.328
	Aluminium alloy	5019	353	39.536	39.536
		6082	353	43.772	43.772

However, there are also disadvantages of welding i.e. welded connections. A few are listed below:

- possible overheating of material
- HAZ weakening
- welding requires skilled labour
- it is difficult to check and test joints
- transport problems due to size of structure
- little space for tolerances, expansion and shrinkage

Apart from the different joining method, the design principle and method application for finding the load path are the same as for mechanical joints.

8.2.1 Load path

The simplicity of a welded connection is reflected by its load path. Whilst for a bolted connected the load path could have a great number of parts by which the load is transferred, the parts making up the load path for a welded connection are usually much fewer. Figure 8.3 illustrates the load path parts for the joint shown in Figure 8.2 by using a fillet weld.

Compared to the load path from Figure 8.2 the welded joint load path parts are reduced to three instead of five. This results in less design work, fewer joint parts and saving in fabrication time. However, further considerations when designing and detailing the joint will be necessary to prevent crevices between the beam and the column. Chapter 9 discusses how to prevent crevices and corrosion by design.

Directions for designing joints are given by the relevant design codes, e.g. Eurocode 9 (EN 1999–1) Clause 8 *Design of joints*.

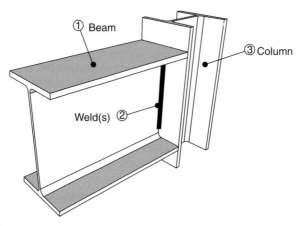

Figure 8.3 | Example of load path for welded joint

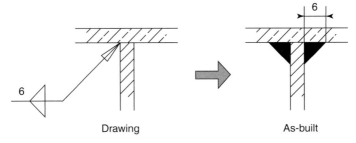

| Drawing | As-built |

Figure 8.4 | Fillet welding: drawing and as–built

8.2.2 Weld types

The most commonly used types of weld for structural applications joining aluminium alloy elements are fillet welds and butt welds. Plug and slot welds are also possible but are more rarely utilised and are then limited to special locations and conditions where fillet or butt welding would not be possible.

8.2.2.1 Fillet welds

Fillet welding is the most obvious choice for welding surfaces or elements at an angle or perpendicular to each element. The fundamental, theoretical weld shape is an isosceles triangle, mostly with a 90° backside angle (see Figure 8.4).

As shown in Figure 8.4 the size and specification of the fillet weld is given by the length of one side of its isosceles triangle, this side is termed the leg length. It is a theoretical dimension as the real weld will penetrate into the parent materials. However, the leg length g gives the minimum required weld size. to design and verify the strength of the weld. The throat thickness a is used to design and verify the strength of the weld. For an isosceles right-angled triangle, the ideal shape of a fillet weld, the throat thickness a can easily be calculated from the leg length g:

$$a = g\sqrt{2}$$

(8.6)

Figure 8.5 shows the terminology of the criteria and dimensions for a fillet weld.

8.2.2.2 Butt welds

Butt welds are used to join elements in line, for example two plates welded together, but can also be adopted for T- and L-shaped joints. Butt welds are more complex than fillet welds, therefore they require more preparation and skill to execute. Butt welds are subdivided in categories and named according their shape, edge preparation of the

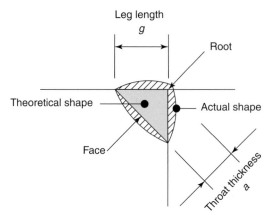

Figure 8.5 | Fillet weld terminology

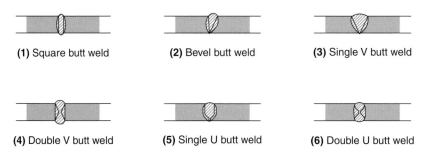

Figure 8.6 | Typical shapes for butt welds

parent material and penetration depth. Typical butt welds are shown in Figure 8.6. Butt welds are also categorised by the penetration depth in full penetration welds and partial penetration welds, an important factor when calculating weld capacities.

The illustrated examples are all full penetration butt welds. It is recommended to use full penetration butt welds for primary structural members and associated connections. This guidance is also expressed in Eurocode 9 (EN 1999–1–1) where partial penetration butt welds should only be used for secondary and non load-bearing members.

As a simple rule of thumb it can be assumed that a well-executed full penetration butt weld should have the same strength as the parent material with the lesser strength. Nevertheless, this would not take into account HAZ weakening, and a connection should always be confirmed by calculations.

When specifying full penetration butt welds it is beneficial to make use of backing, especially for members with greater wall thickness

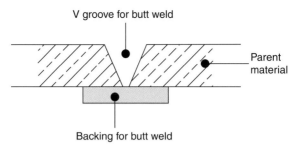

Figure 8.7 | Backing for butt welds

Table 8.5 | Typical weldable aluminium alloys

Weldable aluminium alloys by series

1xxx	3xxx	4xxx	5xxx	6xxx	7xxx
1050A	3003	4015	5005*	6005A*	7020*
1100	3004*		5005A*	6016	
1200	3005*		5049*	6060*	
	3103*		5052*	6061*	
	3105		5083*	6063*	
			5454*	6082*	
			5754*	6106*	
				6262	

* Use with EN 1999–1–1, Tables 3.2 a, b, c

List is not exhaustive, but most common alloys listed

(see Figure 8.7). Due to the possibilities given by extruded sections, backing can already be provided by one of the parent members.

8.2.3   Weldable aluminium alloys

In general, most aluminium alloys are weldable. However, there are a few that are unsuited for welding either due to weld cracking or to excessive heat weakening by the welding process. Table 8.5 lists some of the most common weldable aluminium alloys. The list must not be seen as either exhaustive or complete, thus alloys that are not listed might still be weldable.

8.2.4   Weld capacities

The equations listed below have been employed to determine the weld capacities for aluminium alloy joints. Guidance from Eurocode 9

(EN 1999–1–1) was implemented in deriving these equations. The symbol a was used for the throat thickness of both butt and fillet welds. Equations (8.7)–(8.10) yield the design resistance of welds, limited to the weld alone and condone weakening due to the effects of HAZ.

- for butt welds:

$$F_{Rd\|} = 0.6 \times a \times f_w / \gamma_{Mw} \text{ (kN/mm)} \tag{8.7}$$

$$F_{Rd\perp} = a \times f_w / \gamma_{Mw} \text{ (kN/mm)} \tag{8.8}$$

- for fillet welds:

$$F_{Rd\|} = 0.6 \times a \times f_w / \gamma_{Mw} \text{ (kN/mm)} \tag{8.9}$$

$$F_{Rd\perp} = 0.707 \times a \times f_w / \gamma_{Mw} \text{ (kN/mm)} \tag{8.10}$$

where:

$F_{Rd\|}$ = design resistance parallel to weld
$F_{Rd\perp}$ = design resistance perpendicular to weld
a = throat thickness
f_w = characteristic strength of weld material
γ_{Mw} = partial safety factor for welded joints

The weld capacity of a welded connection is then obtained by multiplying the design resistance by the effective length of the associated weld. In addition to the above, allowance for a HAZ must be made following specific design methods as laid out in Clause 8.3.6.4 and applying the relevant HAZ factors to the design resistances of welded joints. Further information on HAZ weakening can be found in Chapters 2 (Section 2.7) and 7 of this book.

Table 8.6 gives the design resistances for butt and fillet welds using 5356 alloy filler material and most common aluminium alloys worked with in construction. The material properties are taken from EN 1999–1–1, Table 8.8.

Table 8.6 | Weld capacities to EC9

Filler/ alloy	Weid type	Leg (mm)	Throat (mm)	f_w (N/mm²)	γ_{Mw}	$F_{Rd\parallel}$ (kN/mm)	$F_{Rd\perp}$ (kN/mm)
5356/6082	Butt		2	210	1.25	0.202	0.336
			3	210	1.25	0.302	0.504
			4	210	1.25	0.403	0.672
			5	210	1.25	0.504	0.840
			6	210	1.25	0.605	1.008
			7	210	1.25	0.706	1.176
			8	210	1.25	0.806	1.344
			9	210	1.25	0.907	1.512
			10	210	1.25	1.008	1.680
			12	210	1.25	1.210	2.016
	Fillet	3	2.1	210	1.25	0.214	0.252
		4	2.8	210	1.25	0.285	0.336
		5	3.5	210	1.25	0.356	0.420
		6	4.2	210	1.25	0.428	0.504
		7	4.9	210	1.25	0.499	0.588
		8	5.7	210	1.25	0.570	0.672
		9	6.4	210	1.25	0.641	0.756
		10	7.1	210	1.25	0.713	0.840
		12	8.5	210	1.25	0.855	1.008
5356/7020	Butt		2	260	1.25	0.250	0.416
			3	260	1.25	0.374	0.624
			4	260	1.25	0.499	0.832
			5	260	1.25	0.624	1.040
			6	260	1.25	0.749	1.248
			7	260	1.25	0.874	1.456
			8	260	1.25	0.998	1.664
			9	260	1.25	1.123	1.872
			10	260	1.25	1.248	2.080
			12	260	1.25	1.498	2.496
	Fillet	3	2.1	260	1.25	0.265	0.312
		4	2.8	260	1.25	0.353	0.416
		5	3.5	260	1.25	0.441	0.520
		6	4.2	260	1.25	0.529	0.624
		7	4.9	260	1.25	0.618	0.728
		8	5.7	260	1.25	0.706	0.832
		9	6.4	260	1.25	0.794	0.936
		10	7.1	260	1.25	0.882	1.040
		12	8.5	260	1.25	1.059	1.248

8.3   Bonded connections

Adhesively bonded connections have become an alternative to welding and mechanical jointing methods over the past years. As with the other connection techniques, adhesive bonded joints have advantages but also drawbacks. Their advantages are:

- no HAZ weakening
- no cross-section reduction due to holes
- no deformation due to welding
- no fasteners or welds visible = better aesthetic

Their disadvantages are:

- They are limited to shear connections (see Eurocode 9, Annex M)
- careful surface preparation is needed
- shear strength of adhesives only a fraction of that of parent material
- limited lifespan when used in adverse conditions, exposed to heat and submerged in water for prolonged periods

Limited design guidance and criteria for the design and use of bonded connection is given by EN 1999–1-1, Annex M. However, it is best to seek expert advice on this subject. Design and execution of the bonded connections must be undertaken with great care when utilised in primary structures.

9

CORROSION PREVENTION

Introduction

Aluminium alloys are often praised for their excellent resistance to corrosion in most environments. Nevertheless, research carried out on the subject of corrosion of aluminium and aluminium alloys has identified a number of possible corrosion types, where damage to the surface protective layer can occur due to weathering, imperfections and bi-metallic reactions. Knowledge of these corrosion concerns will allow suitable measures to be employed to protect the members from exposure to these attacks.

In principle, methods of preventing corrosion can be subdivided into two categories: prevention by design and prevention by surface treatment. As different as the two methods are, both depend on the correct knowledge of the material, use, environment and exposure of aluminium structural members.

This chapter gives a first insight into corrosion prevention of aluminium and aluminium alloy elements. A case-by-case examination of each element, member or structure is necessary in order to determine which method is best suited to which situation.

Given the current level of research into, and development of, new methods of surface modification rather than simple surface treatments, these new methods, such as laser treatment and ion implantation, could become alternatives to, or perhaps replacements for, standard chemical and coating treatments in the near future.

9.1   Corrosion prevention by design

Structural design often focuses on design criteria such as displacement and structural strength, neglecting the peripheral design considerations associated with corrosion prevention amongst other topics. Aside from the immediate considerations of structural strength, good design must also address the possibility of corrosion over the lifespan of a structure. Premature failure of a structure or individual elements of a structure will have consequences that are as serious as structural strength failure. The difference between the two failures is only in the time required for the corrosion-induced weakening of the material to progress. Therefore, corrosion prevention must start at the design and detailing stage.

Design to prevent corrosion is based on the understanding of the corrosion processes and types, what causes corrosion and how corrosion can be managed by design but also by surface treatment or a combination of both.

Methods of corrosion control include: selecting the most suitable material, applying the most appropriate surface treatment, allowing for drainage and giving the required space for maintenance. However, other factors such as poor fabrication and operating conditions can greatly influence the resistance to corrosion.

This chapter is not intended to be exhaustive but should give plenty of encouragement to further reading and research about the subject of corrosion prevention. The intention is to highlight how the corrosion of aluminium and aluminium alloy elements could be prevented or at least slowed to result in the maximum design life of a structure. It may not be possible to completely eliminate maintenance by good design, yet good design can result in fewer maintenance cycles i.e. longer periods between the necessary maintenance work.

9.1.1 Prevention by good detailing

Proper and careful detailing, part of the structural member design stage will help to prevent corrosion. Simple details, such as member orientation to avoid dirt and water accumulation, sympathetic joining of members or suitable fabrication methods, will all assist with corrosion prevention.

9.1.1.1 Cross-section selection

Figure 9.1 illustrates how first steps in corrosion prevention can be incorporated at the early stage of selecting the cross-section or, in the

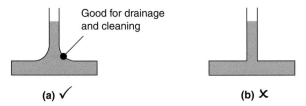

Figure 9.1 | Cross–section design/selection

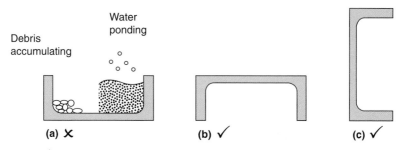

Figure 9.2 | Member orientation

event of purpose-designed extrusion, how small changes can help with maintenance and self-cleansing.

The utilisation of rounded (see Figure 9.1(a)) instead of sharp (see Figure 9.2(b)) corners will prevent dirt accumulations being compressed into the corners. In addition, the curved corner will assist with drainage and provide a better surface for coatings. If possible, sharp internal corners should be avoided.

9.1.1.2 Member orientation

Member orientation for open cross-section shapes can be critical for corrosion prevention. As shown in Figure 9.2(a) a channel section which is open at the top will allow rainwater to build up and also allow dirt and organic material to accumulate, thus fostering possible corrosion.

If, for a particular reason, a channel section is required to be installed with the web facing down, drainage holes can allow water drainage or suitable lid sections can be installed to abolish water and dirt build-up within the channel section (see Figure 9.3).

9.1.1.3 Fabrication of members

Due to the ease and wide availability of extrusion process manufacturing of aluminium and aluminium alloy sections it is not necessary

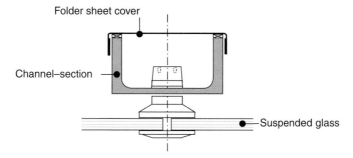

Figure 9.3 | Example for good, preventative design

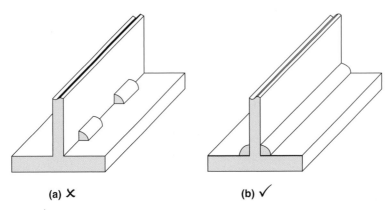

(a) ✗ (b) ✓

Figure 9.4 | Section fabricated by welding: (a) intermittent welding: (b) continuous welding

to fabricate profiles made from plates. However, if only very small quantities of profiles are needed, it might still be more economic to fabricate sections from plates or join other stock cross-sections together. In this process intermittent welding is applied with the aim of limiting welding weakening and overheating. However, non-continuous welding leaves crevices and the detail that is formed is poor from the point of view of corrosion prevention. As shown in Figure 9.4(b), continuous welding will prevent the formation of crevices and consequently give better corrosion protection than the intermittent weld (see Figure 9.4(a)).

Whilst welding is very commonly used to join or fabricate members, it should be avoided if possible. The welding process may cause changes to the metallurgical properties of the materials which, in turn, could reduce the resistance to corrosion. Thus, welding should always be considered carefully and executed with caution.

9.1.2 Avoidance of bi-metallic contact

Galvanic corrosion, also known as bi-metallic corrosion, occurs when aluminium and aluminium alloy elements are in direct contact with other metals which have different electrode potentials. However, an electrolyte is needed to start the ion migration from anode to cathode. This, in turn, leads to the anodic metal corroding at accelerated rates. The galvanic potential of a metal is given by its position in the galvanic series. A simplified extraction of the galvanic series can be found in Section 2.8 of this book.

Galvanic corrosion is severe and can be prevented by: isolating the different metals, the use of sacrificial anodic material, maintaining an electrolyte-free environment i.e. keeping it dry or suitable surface treatment. However, galvanic corrosion can also be controlled

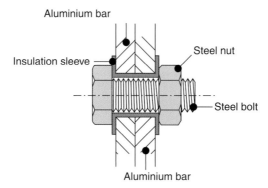

Figure 9.5 | Example of electrical insulation for a bolted joint

by material selection where the materials in contact should be close within the galvanic series.

In summary: galvanic corrosion can be controlled by:

- use of same material if possible
- electrical insulation of metals (see Figure 9.5)
- dry environment (keeping it dry)
- surface treatment (coating)
- use of a sacrificial material that is less noble

9.1.3 Avoidance of crevices

Crevices, that allow crevice corrosion, can form at narrow openings, spaces between metal surfaces or spaces between metal surfaces and non-metal surfaces. Examples of crevices include: flanges, washers, threaded joints and joints with gaps, but also cracks and seams. Further information can be found in Section 2.8 of this book. Crevice corrosion can be limited by:

- reducing contact between metal and non-metal
- applying filler material i.e. mastics to fill crevices
- avoiding sharp corners and edges
- welding joints instead of bolting

Figure 9.6 illustrates a simple method to avoid crevice corrosion by applying a filler material to possible crevices. The use of mastic filler will not completely eliminate crevice corrosion but will minimise the risk of it occurring.

9.1.4 Design for coating

Prior to pre-treatment, i.e. surface preparation for coating, the design or selection of the most suitable section shape can provide additional

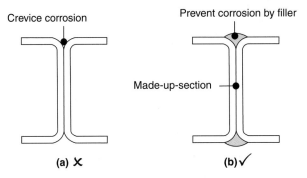

Figure 9.6 | Example of prevention of crevice corrosion

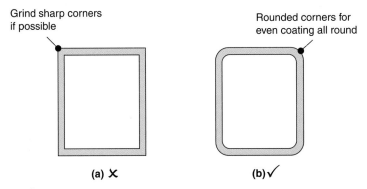

Figure 9.7 | Example of good design for coating

aid for a uniform film thickness of the coating and increased adhesion on a maximised surface area. Achieving a uniform film thickness without weak lines and areas is desirable all round. Sharp edges should be avoided if possible (see Figure 9.7). Stock section shapes with sharp edges can be improved by chamfering/grinding the edges.

9.1.5 Maintenance

Making allowances for maintenance and essential repairs will result in an increased lifespan of a structure or building. Access and space for maintenance and repair work should be considered at the design stage. Good design and ease of maintenance can be achieved by simple measures such as: member spacing, member orientation and ease of access but also safety during maintenance (e.g. using working platforms, ladder restraints and fixing points).

9.2 Corrosion prevention by surface treatment

Although there is a natural protective film on the surface of aluminium and aluminium alloy elements, it is often necessary to provide further protection or simply apply a decorative finish. The raw finish after man-ufacture (i.e. milling, rolling or extrusion) is known as the 'mill finish'. The appearance of this raw finish can vary from a dull greyness to a bright metallic sheen as a result of the properties of the material and the production processes. Many types and methods of surface treatments are available. A few of the most common treatments are listed below.

9.2.1 Cleaning/pre-treatment

As with other materials, aluminium and aluminium alloy elements require careful preparation prior to the application of the surface treatment or finish. The required method i.e. pre-treatment depends on criteria such as: material composition, manufacture, expo-sure and desired finish of the elements. Surface cleaning and pre-treatment should not be neglected, as absent or poor pre-treatment will result in a poor finish and associated problems. Surface clean-ing and pre-treatments can be subdivided into groups based on the method used:

- chemical pre-treatments
- electro-chemical pre-treatments
- mechanical pre-treatments

Typical surface preparations (pre-treatments) irrespective of the above groups are:

- cleaning
- etching
- polishing

9.2.1.1 Cleaning

The cleaning of aluminium elements is part of the chemical pre-treatments. The most common methods are utilising solvents and emul-sions. Cleaning aims to remove not only surface contamination such as dirt and fabrication lubricants, but also protective coatings which were used to prevent corrosion during storage and transport. Methods of cleaning include: wiping, immersion in a bath, and vapour treatment.

9.2.1.2 Etching

Similar to the cleaning process, etching belongs to the group of chemi-cal pre-treatments. Whilst cleaning will only remove surface contami-nations, the etching process, often referred to as chemical etching, will dissolve off the surface layer, oxides and staining. Etching results in a thin, uniform protective oxide film. Etching is often carried out to prepare aluminium and aluminium alloy elements for anodising.

A typical process involves submersion of the parts in a heated caustic soda solution for approximately 10–15 minutes. Etching residues, which are often formed at the surface due to impurities and alloying elements, must be removed by dipping the item briefly into nitric acid.

9.2.1.3 Polishing

Polishing is used in chemical, electro-chemical and mechanical pre-treatments. Therefore, an associated method of polishing can be specified for each group:

- chemical polishing
- electro-polishing
- mechanical polishing

The aim of polishing whether chemical, electro-chemical or mechanical is to prepare a smooth and metallic bright surface. The polished surface can be kept without further surface treatment but can also treated by coating, anodising and plating.

If it is necessary to produce a clean rough surface with good adhesive properties for primer and paint coating, blasting methods are best suited such as:

- sand blasting
- grit blasting
- dry-ice blasting
- shot blasting

The basic process for polishing is the same for all polishing methods. A smooth and clean surface is formed by removing any peaks and projections of uneven and coarse surface areas. However, whilst mechanical polishing is achieved by grinding or blasting, chemical polishing and electro-polishing are dissolution processes where the projecting spots are dissolved leaving a smooth surface as required.

9.2.2 Coatings

After suitable and careful surface pre-treatment aluminium and aluminium alloy elements can be protected and aesthetically enhanced by the process of coating. Coatings available are subdivided in inorganic and organic coatings, where inorganic coatings are often referred to as conversion coatings.

9.2.2.1 Conversion coatings

Inorganic coatings (conversion coatings) are produced by chemical and/or electro-chemical application. This method changes the surface layer of the aluminium and aluminium alloy elements i.e.. the conversion results in an increased resistance to corrosion compared to the natural protective film. Anodising falls within this group and is explained separately.

Conversion coatings are applied not only to enhance corrosion resistance, but also to improve the adhesion of bonding agents or organic coatings. Typical conversion coatings for the use with aluminium and aluminium alloy elements include:

- chromate conversion
- phosphate conversion
- oxide conversion

9.2.2.2 Organic coatings

Organic coatings are the most commonly applied coatings for aluminium and aluminium alloy elements. Again, the surface pre-treatment and selection of the most suitable coating is of importance to give the required result for adhesion to the material and the desired appearance. Organic coatings are widely known and include:

- powder coatings
- primers
- paints and lacquers
- bituminous, silicone and rubber coatings

Organic coatings are usually applied by brushing, dip-coating or spraying. An exception to this is powder coating. The process of powder coating consists of two main parts: first, the powder application; and secondly, the heating/curing process. During the application of the powder, the dry powder is sprayed into the element by an electrostatic spray gun. Due to the electrostatic charge the powder adheres to the object, thus providing a uniform, dry power coverage. When heated the powder melts and flows forming a very smooth and even surface. Compared to wet coatings, powder coating is a solvent-free and environmentally sustainable method of organic coating. Good durability and ultraviolet resistance are desirable for external use.

Organic coatings are often used for aesthetic reasons rather than corrosion protection. Nevertheless, they provide good protection by forming a barrier i.e. an additional layer between the material and its surroundings/exposure to corrosion attack.

9.2.3 Anodising

Aluminium and aluminium alloy elements can be anodised to increase corrosion resistance. The electro-chemical process of anodising produces an oxide film on the material surface, thus enhancing the resistance to corrosion. The oxide film that is formed is transparent and colourless, however, it can be coloured by dying, hard colouring (self-colouring/integral colouring) and electrolytic colouring. A wide variety of colours can be achieved in a modern anodising plant.

Anodising can also be deployed as a surface pre-treatment. Anodised surfaces will provide better adhesion for paint, primers and adhesives than non-anodised elements.

The typical steps for anodising an aluminium element include:

(1) racking
(2) cleaning
(3) anodising
(4) colouring
(5) sealing
(6) un-racking/packing
(7) testing if required

Anodising produces a low maintenance, corrosion resistant, hard wearing but also decorative surface.

9.2.4 Plating

Plating involves depositing a metallic film on the pre-treated surface of aluminium or aluminium alloy elements by electrolytic plating, electroless plating or vapour deposition. Again, the aim is to enhance the existing properties such as corrosion resistance, wear resistance, temperature resistance and appearance, and also to provide additional qualities such as magnetic properties and surface hardness.

Metals that can be plated onto aluminium include:

- cadmium
- chromium
- cobalt
- copper
- gold
- lead
- nickel
- rhodium
- silver
- tin

The typical steps for plating an aluminium element are:

(1) cleaning (degreasing, etching)
(2) de-oxidation

(3)   ion exchange
(4)   plating
(5)   additional plating if required

It is usually more expensive to use plating of aluminium and aluminium alloy elements rather than the other surface treatments. However, the combination of metals which can be used and the method itself achieve excellent results compared to other treatments.

APPENDICES

Appendix 1   Tables of dimensions and properties

Unlike section shapes produced for structural steel applications, where an advanced standardisation has resulted in widely recognised section shapes being available, such standardisation has not taken place to a similar extent for section shapes to be used for structural aluminium applications.

Due to the ease of manufacture and multitude of manufacture facilities an almost inexhaustible number of section shapes are available for sections made from aluminium and aluminium alloys. The following tables list the dimensions, some cross-sections and the structural properties of the most common aluminium-based section shapes.

The illustrated section shapes (see Figure A.1) are only a small selection of the section shapes that are readily available. However, upon request most manufacturers or extruders will produce any possible section shape to suit requirements and functionality.

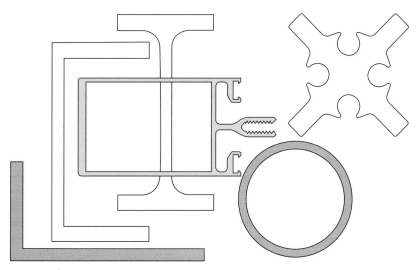

Figure A.1 | Typical section shapes

Table A.1.1 | I- and H-sections

I- and H-sections

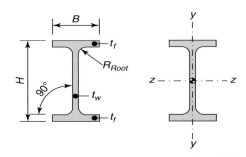

Label	H	B	t_w	t_f	R_{Root}	I_y	W_y	Area
	(mm)	(mm)	(mm)	(mm)	(cm⁴)	(cm³)	(cm³)	(cm²)
60 × 30	60	30	4	6	6.0	31.6	10.5	5.83
75 × 50	75	50	2.5	2.5	2.0	40.4	10.7	4.28
80 × 40	80	40	5	7	7.5	91.6	22.9	9.38
80 × 60	80	60	5	6	5.0	114	28.5	10.8
100 × 50	100	50	6	8	9.0	210	42.1	13.7
100 × 100	100	100	5	5	5.0	260	52.0	14.7
120 × 60	120	60	6	9	9.0	403	67.2	17.6
120 × 120	120	120	5	8	6.0	657	109	24.7
140 × 70	140	70	7	10	10.5	725	104	23.3
160 × 80	160	80	7	11	10.5	1174	147	28.2
160 × 160	160	160	6.5	9.5	5.0	1885	235	39.8
260 × 120	260	120	5	12	7.0	5036	387	41.0

Section shapes with rounded toes are available on request and stock sections with some manufacturers and extruders

Table A.1.2 │ Equal angle sections

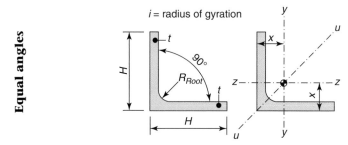

Label	H	t	R_{Root}	x	i	I_y	W_y	Area
	(mm)	(mm)	(mm)	(mm)	(mm)	(cm⁴)	(cm³)	(cm²)
50 × 50 × 4	50	4	4	13.9	15.4	9.28	2.57	3.87
60 × 60 × 5	60	5	5	16.7	18.5	19.9	4.61	5.80
70 × 70 × 6	70	6	6	19.5	21.6	37.9	7.52	8.11
80 × 80 × 6	80	6	6	22.0	24.8	57.4	9.92	9.31
80 × 80 × 8	80	8	8	22.8	24.5	73.9	12.9	12.3
90 × 90 × 8	90	8	8	25.3	27.7	107	16.5	13.9
100 × 100 × 8	100	8	8	27.8	30.9	148	20.5	15.5
100 × 100 × 10	100	10	10	28.5	30.6	180	25.2	19.2
120 × 120 × 6	120	6	6	32.1	37.7	201	22.8	14.1
120 × 120 × 10	120	10	10	33.5	37.1	319	36.9	23.2
125 × 125 × 10	125	10	10	34.7	38.7	363	40.2	24.2
150×150 × 10	150	10	10	41.0	46.7	639	58.6	29.2

Section shapes with rounded toes are available on request and stock sections with some manufacturers and extruders

Table A.1.3 | Unequal angle sections

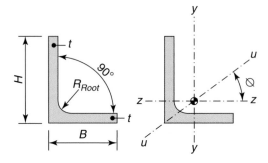

Unequal angles

Label	H	B	t	R_{Root}	Ø	I_y	W_y	Area
	(mm)	(mm)	(mm)	(mm)	(°)	(cm⁴)	(cm³)	(cm²)
50 × 40 × 4	50	40	4	4	32.0	8.65	2.49	3.47
60 × 40 × 4	60	40	4	4	23.8	14.3	3.53	3.87
70 × 50 × 5	70	50	5	5	26.6	28.9	6.04	5.80
75 × 50 × 6	75	50	6	6	23.6	41.2	8.17	7.21
80 × 40 × 6	80	40	6	4	14.6	45.4	8.88	6.87
80 × 50 × 6	80	50	6	6	21.2	49.2	9.24	7.51
90 × 60 × 6	90	60	6	6	23.8	72.6	11.9	8.71
100 × 50 × 6	100	50	6	6	14.8	91.1	14.1	8.71
100 × 65 × 8	100	65	8	6	22.7	128	19.2	12.6
125 × 75 × 9	125	75	9	6	19.9	278	33.6	17.2
130 × 65 × 8	130	65	8	6	14.8	265	31.7	15.0
150 × 75 × 10	150	75	10	6	14.8	505	52.4	21.5

Section shapes with rounded toes are available on request and stock sections with some manufacturers and extruders

Table A.1.4 │ T-sections

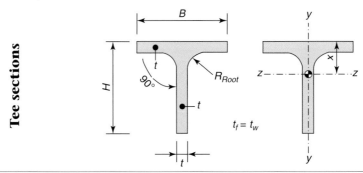

Label	H	B	t	R_{Root}	x	I_y	W_y	Area
	(mm)	(mm)	(mm)	(mm)	(mm)	(cm⁴)	(cm³)	(cm²)
50 × 40 × 3	50	40	3	3	14.8	6.70	1.91	2.64
50 × 50 × 3	50	50	3	3	13.4	7.18	1.96	2.94
60 × 50 × 5	60	50	5	5	17.9	18.9	4.50	5.35
60 × 60 × 5	60	60	5	5	16.6	20.0	4.62	5.85
75 × 50 × 5	75	50	5	5	24.0	35.2	6.91	6.10
80 × 80 × 8	80	80	8	6	22.7	74.0	12.9	12.3
100 × 50 × 10	100	50	10	5	36.9	142	22.5	14.1
100 × 70 × 4	100	70	4	4	30.6	70.4	10.1	6.71
100 × 100 × 10	100	100	10	6	28.5	180	25.2	19.1
120 × 60 × 5	120	60	5	4	41.6	135	17.2	8.82
120 × 120 × 6	120	120	6	4	32.1	201	22.8	14.1
125 × 75 × 8	125	75	8	8	41.5	252	30.2	15.6

Section shapes with rounded toes are available on request and stock sections with some manufacturers and extruders

Table A.1.5 | Channel sections

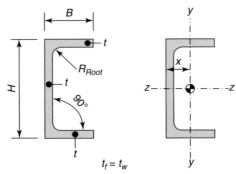

Label	H	B	t	R_{Root}	x	I_y	W_y	Area
	(mm)	(mm)	(mm)	(mm)	(mm)	(cm⁴)	(cm³)	(cm²)
50 × 40 × 4	50	40	4	2.5	13.7	19.5	7.82	4.91
60 × 40 × 4	60	40	4	4	12.8	30.2	10.1	5.34
70 × 45 × 4	70	45	4	4	13.7	56.1	14.9	6.34
80 × 50 × 5	80	50	5	5	15.6	85.9	21.4	8.61
90 × 50 × 5	90	50	5	4	14.9	112	25.1	9.07
100 × 50 × 6	100	50	6	4	14.6	168	33.6	11.3
120 × 55 × 6	120	55	6	4	15.3	279	46.5	13.1
130 × 50 × 5	130	50	5	5	12.6	271	41.7	11.1
140 × 60 × 7	140	60	7	6	16.3	494	70.6	17.3
150 × 50 × 6	150	50	6	4	12.2	445	59.4	14.3
180 × 60 × 5	180	60	5	5	13.8	671	74.6	14.6
200 × 65 × 9	200	65	9	6	16.1	1532	153	28.2

Section shapes with rounded toes are available on request and stock sections with some manufacturers and extruders

Table A.1.6 | Z-sections

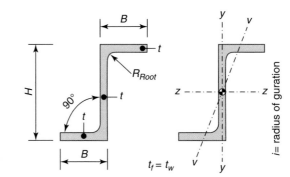

Label	H	B	t	R_{Root}	i_{min}	I_y	W_y	Area
	(mm)	(mm)	(mm)	(mm)	(mm)	(cm⁴)	(cm³)	(cm²)
30 × 60 × 30	60	30	3	3	11.5	18.8	6.28	3.45
40 × 80 × 40	80	40	3	3	15.6	46.2	11.5	4.65
50 × 100 × 50	100	50	4	4	19.5	119	23.9	7.74
60 × 120 × 60	120	60	5	5	23.4	257	42.8	11.6
70 × 140 × 70	140	70	6	6	27.2	488	69.7	16.2
80 × 160 × 80	160	80	8	8	30.9	952	119	24.6
90 × 180 × 90	180	90	8	8	34.9	1377	153	27.8
100 × 200 × 100	200	100	10	10	38.6	2325	232	38.4
120 × 240 × 120	240	120	10	10	46.7	4113	342	46.4

Section shapes with rounded toes are available on request

Table A.1.7 | Circular hollow sections: circular tubes

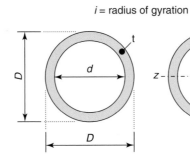
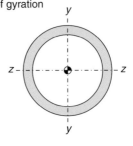

i = radius of gyration

Label	D	t	d	i	I_y	W_y	Area
	(mm)	(mm)	(mm)	(mm)	(cm⁴)	(cm³)	(cm²)
∅60 × 5	60	5	50	19.5	32.9	10.9	8.64
∅70 × 5	70	5	60	23.0	54.2	15.5	10.2
∅80 × 5	80	5	70	26.5	83.2	20.8	11.8
∅90 × 5	90	5	80	30.1	121	26.9	13.4
∅90 × 10	90	10	70	28.5	204	45.4	25.1
∅100 × 5	100	5	90	33.6	168	33.7	14.9
∅100 × 10	100	10	80	32.0	289	57.9	28.3
∅110 × 5	110	5	100	37.1	227	41.4	16.5
∅120 × 5	120	5	110	40.7	299	49.8	18.1
∅130 × 4	130	4	122	44.5	314	48.4	15.8
∅140 × 3	140	3	134	48.4	303	43.3	12.9
∅150 × 5	150	5	140	51.3	599	79.9	22.7
∅160 × 5	160	5	150	54.8	731	91.5	24.3

Circular tubes

Table A.1.8 | Square hollow sections: square tubes

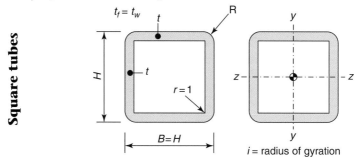

i = radius of gyration

Label	H	t	R	i	I_y	W_y	Area
	(mm)	(mm)	(mm)	(mm)	(cm⁴)	(cm³)	(cm²)
50 × 50 × 4	50	4	5	18.6	24.8	9.96	7.14
60 × 60 × 5	60	5	6	22.3	53.2	17.7	10.6
70 × 70 × 5	70	5	6	26.4	88.4	25.2	12.7
80 × 80 × 6	80	6	7	30.0	156	39.1	17.3
90 × 90 × 4	90	4	5	34.9	165	36.8	13.5
90 × 90 × 6	90	6	7	34.1	230	51.1	19.7
100 × 100 × 5	100	5	6	38.6	279	55.8	18.6
100 × 100 × 6	100	6	7	38.2	323	64.6	22.1
120 × 120 × 5	120	5	6	46.8	496	82.8	22.7
120 × 120 × 7	120	7	8	45.9	656	109	31.1
140 × 140 × 5	140	5	6	54.9	806	115	26.7
150 × 150 × 5	150	5	6	59.0	1000	133	28.7
160 × 160 × 9	160	9	10	61.4	2018	252	53.4

Table A.1.9 | Rectangular hollow sections: rectangular tubes

Label	H	B	t	R	i_z	I_y	W_y	Area
	(mm)	(mm)	(mm)	(mm)	(mm)	(cm⁴)	(cm³)	(cm²)
$50 \times 30 \times 3$	50	30	3	4	11.7	13.4	5.36	4.30
$50 \times 40 \times 3$	50	40	3	4	15.4	16.7	6.69	4.90
$60 \times 30 \times 3$	60	30	3	4	11.9	21.3	7.11	4.90
$60 \times 50 \times 4$	60	50	4	5	19.1	38.9	12.9	7.94
$70 \times 50 \times 4$	70	50	4	5	19.5	56.9	16.2	8.74
$80 \times 50 \times 4$	80	50	4	5	19.8	79.3	19.8	9.54
$100 \times 50 \times 3$	100	50	3	4	20.7	108	21.7	8.50
$100 \times 60 \times 3$	100	60	3	4	24.6	122	24.5	9.10
$120 \times 60 \times 4$	120	60	4	5	24.7	247	41.2	13.5
$150 \times 50 \times 5$	150	50	5	6	20.6	474	63.2	18.6
$180 \times 80 \times 5$	180	80	5	6	33.5	997	110	24.6
$200 \times 100 \times 5$	200	100	5	6	41.9	1491	149	28.7
$220 \times 100 \times 8$	220	100	8	9	41.0	2845	258	47.9

Appendix 2   Types of butt weld joint

Symbol	Figure	t (mm)	b (mm)	h (mm)	α (°)
‖		≤ 16	≤ 3	-	-
V		4–20	0–2	≤ 2	50–100
Y		> 6	3–7	2–4	15–30
Y		> 10	≤ 4	2–6	50–70
U		> 10	≤ 1	2–4	≤ 10
X		> 10	≤ 2	3–4	50–70

BIBLIOGRAPHY

Aluminum Association Inc (AA), 2006. *Registration Record Series, Teal Sheets, International Alloy Designations and Chemical Composition Limits for Wrought Alumium and Wrought Aluminum Alloys*. AA, Arlington, VA, USA.

Aluminium Federation Limited (ALFED), 1993. *The Properties of Aluminium and its Alloys*, 9th edition. ALFED, Birmingham, UK.

British Standards Institution (BSI), BS EN 573–1:2004[2], *Aluminium and aluminium alloys, Chemical composition and form of wrought products – Part 1: Numerical designation system*. BSI, London, UK.

British Standards Institution (BSI), BS EN 1780–1:2002, *Aluminium and aluminium alloys, Designation of alloyed aluminium ingots for remelting, master alloys and castings – Part 1: Numerical designation system*. BSI, London, UK.

British Standards Institution (BSI), BS EN 1990:2002, *Eurocode – Basis of structural design*. BSI, London, UK.

British Standards Institution (BSI), BS 8118–1:1991, *Structural use of aluminium, Part 1: Code of practice for design*. BSI, London, UK.

European Committee for Standardization (CEN), EN 1999–1-1:2007, *Eurocode 9 – Design of aluminium structures – Part 1: General structural rules*. CEN, Brussels, Belgium.

Training in Aluminium Application Technologies (TALAT), 1999. *TALAT Lectures – The TALAT CD-ROM 2.0*. European Aluminium Association, Brussels, Belgium.

INDEX